ENERGY POLICIES, POLITICS AND PRICES

A STRATEGIC EVALUATION OF ENERGY SECURITY IN THE EASTERN MEDITERRANEAN

ENERGY POLICIES, POLITICS AND PRICES

Additional books and e-books in this series can be found on Nova's website under the Series tab.

ENERGY POLICIES, POLITICS AND PRICES

A STRATEGIC EVALUATION OF ENERGY SECURITY IN THE EASTERN MEDITERRANEAN

MICHAIL CHALARIS

nova science publishers
New York

Library of Congress Cataloging-in-Publication Data

ISBN: 978-1-53619-273-5

Published by Nova Science Publishers, Inc. † New York

CONTENTS

PREFACE

'Man pursues Values through Institutions on Resources.'
(Lasswell 1969, 17 [emphasis removed])

The Mediterranean Sea is the most important regional sea on the planet (Tsaltas, 2006). Along with the coastal areas surrounding it, it ranks first among the most important closed regional geopolitical systems since ancient times, with the presence of many important states and a population made up of many nations representing many and different cultures. Geopolitically speaking, it is the most neuralgic, growth point in global shipping and controls, as the ultimate geographical link, the interaction between the vast majority of independent UN member states (Tsaltas, Bourtzis and Rodotheatos, 2009).

Over the years, this area has been identified as one of the most strategic areas in the world. It essentially focuses on uniting geopolitically the Eurasian and the African continents, with the famous red Gulbekian Circle, which includes the countries producing and transporting oil, from as far as East Africa. The Mediterranean Sea is a sea with high salinity due to high evaporation rates and low river flows. It has limited water exchange with the Atlantic Ocean and the Black Sea. It represents the largest biodiversity hot spot in Europe.

The Mediterranean is also considered a semi-enclosed sea, in the sense that, on the west, it is connected to the Atlantic Ocean via the Strait of

Gibraltar and on the east, it is connected to the Red Sea via the Suez Canal. Along with Propontis, the Black Sea and the Sea of Azov, which according to French geographers, are considered its internal parts, It occupies an area of 2,966,000 sq.km or about 800,000 square miles. Of course, one should take into consideration that English geographers don't share this view as they consider these parts unrelated to the Mediterranean Sea, a view that has prevailed over the years, as more practical. Its maximum length, from Gibraltar to the coast of Syria, is 2,100 nm, or 3,860 km, its maximum width is 1,800 km, and its maximum depth is 5,120 meters 62 miles southwest of Cape Tainaro. The average width is 600 km and the average depth is 1,500 meters. In general, the Mediterranean Sea is the largest semi-enclosed sea on Earth and looks like a lake (Tsaltas, Kladi - Efstathopoulou, 2009).

The Mediterranean Sea is a sea of great trade and important maritime strategies, thus uniting different peoples, cultures, religions and worlds. It forms a special geopolitical landscape with states that, diverse in size and shape, are bathed by its waters between three major peninsulas, the Balkan in the east, the Iberian in the west and the Italian in the middle.

The Eastern Mediterranean region can be considered a regional sub-system and is comprised today of nine states, namely Cyprus, Egypt, Greece, Israel, Jordan, Lebanon, Libya, the Palestinian Authority, Syria (to the extent that it can still be treated as a unitary/single actor) Italy and Turkey. This was not the case during the Cold War when the Eastern Mediterranean was thought to include only Greece, Turkey, and the Republic of Cyprus. Today, the Eastern Mediterranean has taken on the grim role the Balkans played in the early 20th century.

Strategy, therefore, is based on the effective control that coastal countries have over these crossings, which are the links in a chain that essentially connects the opposite continental coast. The Straits of the Dardanelles and the Bosphorus, but also the Straits of Kythera and Karpathos (Tsaltas and Deniozos, 2007), east and west of Crete, are a tangible example of this strategic vision of the region at its sensitive south-eastern end.

At the same time, the geo-economic and political parameters that compose, thus complementing the general geo-strategic position of each coastal or island state of the Mediterranean Sea, identifies the serious, often tricky, problems in their vicinity.

The economic and technological development in the field of soft security of the region therefore influences the fair resolution of any interstate disputes, to the extent that it would allow any well-meaning analyst to identify both positive and legitimate proposals and scenarios of cooperation that they would easily overcome any obstacles presented in the field of strict military demonstration in defending territorial integrity (Chalta, 2000).

Moreover, the Eastern Mediterranean is characterized by a unified historical background, bearing elements of a distinct regional subsystem representing the border between East and West as well as the Western world limit (Litsas and Tziampiris, 2015). Stability and security in the contemporary Eastern Mediterranean are drastically affected by an emerging regional complex, consisting of state and non-state entities in search of security. Evidently, regional dynamics and challenges in the Eastern Mediterranean have a direct effect on Europe's security as a whole. Currently, Eastern Mediterranean faces many challenges, being in the spotlight of most regional players due to the migratory flows and the energy developments (Giannakopoulos, 2016; Dokos, 2016).

Energy security remains, "an essential element in national power, a major factor in world economies, a critical focus for war and conflict, and a decisive force in international affairs." Therefore, energy discoveries in the Eastern Mediterranean have generated fresh interest in their potential impact on existing regional disputes and power constellations. The Eastern Mediterranean appears as a separate 'new' region in geo-economic understanding and not as merely an extension of the Middle East (Stivachtis 2019).

The aim of this study is to discuss energy developments and to identify the energy issues that affect the situation in the Eastern Mediterranean region. The individual objectives are the recognition of energy actions and resources, forms of energy, energy security and in general, all energy-related actions mentioned in the specific area. Special attention has been given to recommendations for increasing the EU's capacity for constructive regional intervention.

In order to discuss the main research question and its sub-questions, existing models and theories need to be issued at first. The main research question implicitly contains the theoretical dimensions of energy security, geopolitics, and touches upon the dimensions of International Relations theory (IR-Theory). Further, the choice for these theoretical assumptions is motivated by the fact, that the concepts complete each other and contribute to the answer of each sub-question. Thus, the different theoretical dimensions should not be regarded as separate theories, which will be later instrumentalized to evaluate the Eastern Mediterranean energy security system. Instead, they should build one theoretical framework, stressing the multiple dimensions of energy security, which cannot be addressed by one existing theory. This applies especially to the contemporary demand for considering energy politics in the context of geopolitics and international relations.

For the present research, primary, secondary and tertiary literature sources had been used. Tertiary sources were mainly employed to get an overall understanding of the historic background and composition of Eastern Mediterranean countries' energy policy. Secondary sources were used to identify key actors in implementing and coordinating energy policy in the Eastern Mediterranean as well as to understand the components of the energy mix and complex legal framework and infrastructure related to it.

As primary sources, the author consulted the webpages on the website of the International Energy Agency, International Renewable Energy Agency (IRENA), the European Commission, EU Energy Law, EU Energy Policy Blog, European Energy Security Strategy, United Nations Environment Programme/Mediterranean Action Plan, U.S. Department of Energy, American Petroleum Institute etc. Academic works by Borchert and Foster, Lopez et al., Mc Cormick, Shieldl et al., Stampolis and Tsaltas were of particular importance while formulating the theoretical framework of the present study. Studies by Balodimos, Bornstein, Chorey, Elkind,

Kontos, Tsakiris and some other scholars were taken as the basis for the analysis of the Energy Resources and Energy Security in the East Mediterranean.

As the topic of the present research lies in the field of social sciences, the author employed the tools and instruments of qualitative method, namely data collection, document analysis and single case study. Collected data were used to briefly outline the origins of the energy policy of the countries belonging to Eastern Mediterranean region, to provide definitions of the key terms, to define energy policy in general and energy policy of EU in particular, by covering such topics as its objectives, energy mix, and energy security strategy. Especially this book will first briefly consider what is meant by energy security. Following an overview of security issues in the Mediterranean in general, and a discussion of the EU's increasing interest in the region, the book will examine energy relationships in the Mediterranean Sea with separate analyses on trade in crude oil, natural gas, and LNG. The potentially complicating role of the USA and Russia in the region will be addressed. Concerns over terrorist attacks on energy infrastructure in the Mediterranean, and, within this context, the heightened interest of NATO on issues of energy security, will be also discussed.

Book on a strategic evaluation of energy security in the eastern Mediterranean will consist of seven chapters: preface, five body chapters, conclusion, and references.

The first body chapter presents a general presentation of the topic. The second body chapter reports on energy resources, and it aims to clarify concepts such as energy, energy resources. Furthermore, it focuses on energy security and its basic principles. It presents energy security, its importance, supply disruption, and finally energy crises.

The third body chapter is about energy and security. It aims to study energy and security by focusing on the concept of energy security, the factors that affect energy security, the "4As" system, the policy of differentiation, the unbalanced factors and energy security, the energy security of the infrastructure and in the maritime dimension of the energy security. Furthermore, this chapter acknowledges the fact that energy security is understood not just as an economic but also as a geopolitical and International Relation (IR)-issue.

The fourth body chapter deals with the European Union's interests, strategy and policy regarding the Eastern Mediterranean region. The European Energy and Security Policy Framework will be analyzed.

The fifth body chapter will provide the main findings based on the results in the area we are studying and will present the East Mediterranean energy changes map and data on geopolitical chessboard. In particular, it develops European energy security, the Russia-Iran-Turkey subsystem, the Azerbaijan-Turkey geopolitical axis, the Israel-Cyprus-Greece, the Egypt-Cyprus-Greece geopolitical axes, the US superpower system, the Russian superpower, collisions and the geopolitical "black hole" and finally analyzes the role of Greece in strengthening national and European energy security. Moreover, in the sixth chapter, the conclusions will encompass all findings and summaries.

Finally, we hope that this book will be a useful tool a) to scholars and students b) for researchers in the fields of Energy Economics, Policy, and Security, Energy Law, Business, Regulation and Policy, Geopolitics, who aim to have a better understanding of the current trends or research in the relevant fields, c) for professionals in EU politics and foreign policy, energy policy and security, and more broadly to security studies, European politics and international relations and newcomers in the profession of energy security and d) for policymakers who intend to apply the collective knowledge included into this volume into policy and decision -making.

List of Abbreviations and Acronyms

BP	The British Petroleum Company plc and BP Amoco plc is a multinational oil and gas company headquartered in London, England
bcm	Billion cubic metres of natural gas or cubic kilometer of natural gas is a measure of natural gas production and trade
CIS	Commonwealth of Independent States
EC	European Commission
ECT	Energy Charter Treaty
EEZ	Exclusive Economic Zones
EIS	Energy Infrastructure
ETIs	Energy Transmission Infrastructures
EL.PE. SA	Hellenic Petroleum S.A.
EU	European Union
EURATOM	The European Atomic Energy Community
FIGRA	French Geophysical Institute
4As	Availability, Accessibility, Affordability, Acceptability
Germany RWE	Rheinisch-Westfälisches Elektrizitätswerk – Germany Electricity Company
GHGs	Green House Gasses Emission

ICT	Information and Communication Technology
IEA	International Energy Agency
IESEE	Energy Institute for Southeast Europe
IGME/ HSGME	Hellenic Survey of Geology & Mineral Exploration
IODP	International Ocean Discovery Program
JCPOA	The Joint Comprehensive Plan of Action known commonly as the Iran nuclear deal or Iran deal, is an agreement on the Iranian nuclear program reached in Vienna on July 14, 2015, between Iran and the five permanent members of the Unitcd Nations Security Council—China, France, Russia, United Kingdom, United States—plus Germany together with the European Union.
LNG	Liquefied Natural Gas
NATO	North Atlantic Treaty Organization
NESC	Hellenic National Energy Strategy Council
OPEC	Organization of the Petroleum Exporting Countries
PPC SA	Hellenic Public Electricity Company SA
RES	Renewable Energy Sources
TAP	Trans Adriatic Pipeline
UN	United Nations
USA/US	United States of America

Chapter 1

An Overview of Eastern Mediterranean, Middle East and Energy Issues

"Theory is useful; it enables, it helps us better to understand what we already know, intuitively, in the first place." (Abdulhamidova, 2009, 17)

Energy now, as work formerly, has become something that individuals and societies need. It is a symbol that fits our age, the symbol of that which is both abundant and scarce. (Ivan Illich, 2009, 13)

1.1. Energy Fields in the Eastern Mediterranean

The Eastern Mediterranean region is becoming an important area in the energy map, while it is also a gateway for natural gas to European markets by changing the balances on the geopolitical chessboard. Turkey and the EU face common challenges in the energy field, as they are both heavily dependent on energy imports and have their own suppliers. At the same time, the natural gas pipelines developed, such as Talap and TAP. to transfer natural gas from Azerbaijan's fields, generate new facts. Greece and Turkey are in the right geostrategic position to become the gateway to Europe's markets and it is in the common interest of both countries to work together on energy issues (Sartori, Lorenzo & Paceviciute, 2016).

Firstly, let's repeat some fundamental understanding. Energy is used everywhere. Indeed, we need energy for transport, heating, lighting, production processes. Even the service sector is dependent on energy: how to make a computer work without electricity, how to travel for a business meeting without petroleum, how to communicate without electricity? Therefore, energy is used everywhere and is included in the price of all services and goods we buy. Consequently, it's obvious that every fluctuation of energy price is depicted on the price of goods and services. Thus, no one can doubt the importance of Energy into the Economy.

Considering those, there is a need for a review of the energy policy since it is estimated that energy consumption will increase in the following years/decades due to population growth. At the same time, gas will play an important role, especially for countries with lignite dependence as it is expected to be reduced. In relation to oil products trading, Greece is the largest exporter of liquid and refining products in the Mediterranean. The comparative advantage of Greece oil-refining industry is that it has the know-how and the qualified staff to meet the modern challenges. It should also be mentioned that the energy map of Greece is changing with projects such as the TAP pipeline and the vertical connection of Alexandroupolis. The recent crisis between Russia and Ukraine over the transit of Russian gas through Ukraine to Europe has led to the creation of alternative trade routes such as the TAP and Nordstream 2 (Shields et al., 2011).

Natural gas may be considered as the fuel of the future. Among natural gas main providers, Iran, in order to remain a competitive supplier and exporter of energy, it will need to invest in its mining plants to reduce operating costs and become more competitive. Greece can become a gateway for natural gas to Europe.

Moreover, almost 50% of the proven oil reserves and 46% of oil production in the African continent comes from African countries in the southern Mediterranean region. In addition, 58% of the proven natural gas reserves and 79% of the African continent's production comes from countries in the southern Mediterranean region. The top producing countries are Algeria (natural gas) and Libya (oil), both members of OPEC, with Egypt as a rising power due to the recent discovery of natural gas deposits in the Nile Delta. The EU, due to its geographical proximity, is the main export route for those commodities 58% for oil and 85% for natural gas. Libya's total gas production and 96.5% of Algeria's natural gas production are channeled to the EU (Vakulchuk Overland & Scholten, 2020).

Egypt, on the other hand, is channeling its exports mainly to neighboring countries (e.g., Israel). Due to its geographical location and the absence of onshore pipelines, Egypt chooses to export the natural gas it produces in the form of Liquefied Natural Gas (LNG). In general, LNG is the prevailing form for exporting the natural gas produced in the southern Mediterranean countries. Algeria is one of the world's leading LNG exporters with major destinations in Europe and Asia. In addition to LNG facilities in Egypt, Libya and Algeria, there are also several natural gas terminal stations (regasification plants) on the coast of EU Member States (North Mediterranean) including Spain (Huelva-Cartagena-Barcelona), France (Fos, Comptoir de Bretagne), Italy (La Spezia) and Greece (Revithousa). Turkey also has a natural gas regasification plant in the Sea of Marmara (Ereglisi) (Horden & Purcell, 2019).

The distribution of natural gas from the Mediterranean countries to the EU countries is uneven. In the countries of northern and eastern Europe, natural gas consumption from the Mediterranean is low or non-existent as these countries depend on Russian natural gas imports. Dependence on natural gas imported from Mediterranean countries is higher in southern European countries, with Spain being the EU member state importing the largest amounts of Egyptian natural gas, followed by France, with imports from Algeria mainly, and then Portugal and Italy. Algeria is the most important natural gas exporter in the Southern Mediterranean region with reports indicating that it delivered 41 bcm of gas to the EU in 2011. In particular, Algeria accounts for 13% of EU gas imports and if we look at destination exports, Italy receives 44% of Algerian exports, Spain 26%,

France 10%, Turkey 8%, and also Portugal receives 4 % of Algerian natural gas exports (T,óth & Sebestyén, 2019).

Hydrocarbons are the main source of energy consumption in the southern Mediterranean countries, while the natural gas share in their energy mix is steadily increaded. Apart from Libya, Algeria, Egypt, the rest of the Southern Mediterranean countires is heavily dependent on hydrocarbon imports. Due to demographic growth, energy demand and in particular natural gas demand are expected to increase in the future, but asymmetries in demand and consumption levels are expected between the North (EU Member States) and the South Mediterranean states (Balodimos, 2016).

The region of the Eastern Mediterranean and the Middle East has a special geopolitical significance, due to its unique characteristics. Among others, it is the cradle of the Islamic world and culture, it has the world's largest oil fields, it controls some of the main international trade routes, its territories are crossed by gas pipelines, In addition, it is close to big countries like Russia, China or India. The fierce competition for control and exploitation of oil, economic interests, religious differences and local territorial claims have made this area turbulent, perhaps the most volatile part of the planet. The two-pronged problem of the Palestine, with its conflicts with Israel, the invasion of Iraq in Kuwait, which forced the involvement of the majority of the western world, the intractable Cyprus and Kurdish issues - the Greek-Turkish conflict, are indications but important wounds of modern history in the region, with great and varied extensions. Against this background, Greece as located in the western end of the Eastern Mediterranean basin is called to play its role.

Greece, and its geopolitical continuation the Republic of Cyprus, with which Israel maintains excellent relations for reasons of mutual interest, is seen as "the bridge" to the Western world, especially to the European Union. At the same time, Greece is important for Israel as a landlocked country in the spread of Islamic fundamentalism in the Balkans. From a systemic point of view, the Greece-Israel dipole is part of a geopolitical complex with four ends: Greece-Cyprus-Israel-Turkey. This square intergovernmental subsystem of interactive power processes is first integrated into the broader geopolitical system of the wider Middle East and then into the wider Mediterranean Transnistria, which includes as actors the United States, Russia and the CIS (Commonwealth of Independent States).

In fact, Eastern Mediterranean,has been at the level of energy development, , extremely interesting for several decades, culminating during the last one. The region is gradually becoming an important producer and supplier of energy, mainly natural gas, but also oil, while at the same time it is an energy market with many peculiarities. The search, extraction, production and transportation of hydrocarbons, with all the related processes, will affect significantly the relations among the states of the Eastern Mediterranean, but also with the other states of the international system. Nowadays, despite the technological developments in the field of energy and exploitation of hydrocarbon deposits, their reduction worldwide and the simultaneous increase of the needs of modern society, due to the increase in consumption of hydrocarbons and their products, add another major issue on the agenda of international relations (Tziarras, 2017).

The oil and gas pipelines designed to meet the energy needs of countries of West Europe have to overpass the region of Southeast Europe, if western European countries would prefer to avoid Russia. It is noted that Russia has developed piping infrastructure reaching as far as Eastern Europe, as well as influence in the former Soviet republics. The rich regions of the Middle East and the former Soviet republics, as well as Russia, have brought the countries of Southeast Europe, including Greece, back to the forefront of the attention of the major powers, since they are seen as main energy corridors. The plans of the pipelines, depending on the strength that supports them, form a scenario that shows the influences of the countries in the region (Filis, 2018).

To conclude, the international politics of the Eastern Mediterranean have witnessed a substantial transformation in recent years due to the newly founded energy reserves. In the new power-constellation both regional and non-regional actors have been agonising to increase its power in the region, as the export of existing or potential future energy supplies from the region affects not only the economic security of the surrounding states but also that of major global actors. Based on the newly founded natural resources, one very important new element in this confrontation-complex evolved around regional states' ability to sustain or to diversify their supply of energy. Indeed, at first, these discoveries are an opportunity to decrease the regional energy dependence and an opportunity for economic growth for the energy producing countries. On the other hand the new Eastern Mediterranean hydrocarbon resources could be an important supplementing energy provider for the ever-increasing energy need of the EU. Greece and Cyprus are member states of the European Union and any future energy resources trade between the EU and the Eastern Mediterranean might increase their influence and bargaining leverage inside the decision-making bodies of the Union, while other Eastern Mediterranean countries have a long tradition of bilateral or multilateral relations with the EU and its member states. So far the Eastern Mediterranean is a significant route for EU's natural gas and oil imports, as through the region is trafficked approximately 35% of EU's natural gas consumption and 50% of its oil consumption (Szoke, 2016). Therefore many analysts began underlining the common economic imperatives emanating from a successful exploitation from those energy reserves, a procedure called monetisation, labelling it as a driving force that could ease the existent tensions and rivalries between Lebanese, Syrians, Israelis, Turks, Greeks, Greek-Cypriots and the Turkish Cypriots forcing them to collaborate in order to share economic benefits and revenues. As a matter of fact, theoretically, energy cooperation and interdependence could cultivate an analogous political cooperation, thereby improving overall relations between regional and non-regional actors (i.e., USA, European Union and Russian Federation) who are also in a mediate or immediate way involved in the East Med interstate affairs. In fact, however, the prevailing geopolitical and economic circumstances locally, regionally and globally do not seem to be favourable for such a scenario.

1.2. Energy Fields in the Middle East

1.2.1. The Role of the Middle East in Energy Supply Worldwide

The accumulation of such a large number of hydrocarbons worldwide in the Middle East contributes to a number of economic and national security problems affecting the region and the world at large. The region is full of deep ethnotic and political tensions, terrorism, corruption and authoritarianism. In addition, there are problems that have no solution and will not directly affect energy supply from the Middle East, including protectionism, lack of investment, unresolved border disputes and growing uncertainty about the political stability of basic energy producers such as Saudi Arabia, Iran and Iraq. These problems are likely to intensify as oil demand increases. The problems of the region will undoubtedly affect not only the global economy and security, but also the consumption attitudes and policies of nations towards the producers of the region as well as among themselves (Kiriyama & Kajikawa, 2014).

The only way consumers can control the influence of the region is to put their collective weight together to act in a unified way to offset OPEC by turning to alternatives while at the same time bringing the cartel to policies that contribute in energy security and are needed to reduce oil prices (Luft, 2009).

Throughout the 19th century, nearly half of the world's crude oil supply came from oilfields surrounding the city of Baku. At that time, oil provided only 4% of world energy, giving the Caspian region little strategic advantage on the international stage. However, as the global economy began a steep path of growth, oil dependency increased significantly. Today, oil supplies about 40% of world energy and 95% of transport energy. As a result, those holding the lion's share of the reserves of this valuable energy source are at the heart of the global economy and their influence is increasing (Yergin, 2011).

Since the 1930s the Middle East has emerged as the most important source of energy in the world and the key to the stability of the world economy. This troubled region now produces 37% of the world's oil and

18% of its gas. In terms of reserves, the Persian Gulf is the leading source as it produces 65% of world oil reserves and 45% of natural gas reserves. The Middle East also controls a significant proportion of hydrocarbons that have not yet been explored. According to the US Geographic Survey over 50% of unexplored oil reserves and 30% of natural gas are concentrated mainly in Saudi Arabia, Iran, Iraq, Kuwait, UAE and Libya (Luft, 2009).

The accumulation of such a high amount of hydrocarbons in the world at this geographical location means that as the modern economy depends on oil and gas supplies, the Middle East will play a key role in global politics and economics. Most of the countries in the world are heavily dependent on Persian Gulf oil. In 2006, the Middle East provided 22% of US imports, 36% of OECD countries in Europe, 40% of China, 60% of India, and 80% of Japan and South Korea. Even oil-rich Canada depends on the Middle East since 45% of Canada's oil imports come from the region (Yuanita et al., 2016).

By excluding a major technological transformation, global dependence on the Middle East will increase. According to the International Energy Agency, between now and 2030, global oil consumption will increase by about 60%. Transmission will be the fastest growing oil sector. By 2030, the number of cars will increase to over 1.25 billion from about 700 million today. Consequently, global gasoline consumption could double. The two countries with the highest rate of growth in oil use are China and India, whose populations together represent one third of humanity. Over the next two decades, oil consumption in China is expected to increase at a rate of 7.5% per annum and in India by 5.5% (compared to 1-3% growth for industrialized countries). As a result, by 2030 Asia will import 80% of its total oil needs and 80% of this will come from the Persian Gulf (Luft, 2009).

The reason that the share of the Persian Gulf countries in the global energy pie is likely to increase is not only related to geology but also to resource management. While non-Middle East countries are pumping full speed, Middle Eastern producers, many of whom are members of the Organization of Petroleum Exporting Countries (OPEC), insist on a quota. This means that oil outside OPEC runs out almost twice as fast as OPEC.

Exxon Mobil Corporation has estimated that production outside OPEC, which includes Russia and West Africa, will peak within a decade, making recoverable oil remaining outside the Middle East scarcer (Klare, 2015).

On the other hand, the ratio of reserves to production among the Gulf producers ranges between 80 and 100 years, allowing these countries to remain decades after their competitors have exhausted their reserves. This is likely to lead to global dependence from the region on an unprecedented scale with significant implications for global security and the global economy (Luft, 2009).

1.2.2. The US Role in the Energy Issue

One of the countries directly dependent on energy in the Middle East is China. This is because it has taken full advantage of its own oilfields, which is why imports are increasingly expected to reach much higher rates in the coming years. This is why China has sought to develop diplomatic and economic relations with the Middle East to ensure continuity of supply (Yetiv & Lu, 2007).

China's energy initiatives in the Middle East are generally welcomed in the region. Both Iran and Saudi Arabia responded positively to Chinese intentions. For Iran, the political and strategic advantages of cultivating closer ties with China are obvious. As Tehran is subject to increased international pressure on its nuclear activities, the support of a permanent member of both the UN Security Council and the International Atomic Energy Agency (IAEA) Board provides the necessary international political coverage. Given China's history of supplying weapons and sensitive military technology to Iran, Tehran is certain that Beijing could again play such a role (Leverett & Bader, 2005).

The events in Ukraine have brought about a sharp relief from US's growing energy independence and its resulting geopolitical influence. Almost immediately after Moscow's invasion to Crimea, leaders on both sides of the Atlantic began calling on Washington to strengthen its allies and weaken Russia's dominance by deepening its energy ties with Europe.

In a letter sent to Parliament Preseident John Boehner and Senate Majority Leader Harry Reid, ambassadors from Poland, Hungary, the Czech Republic and Slovakia called on the United States to step up its gas competition to boost energy security in Europe. President Obama reiterated his call stating that energy is obviously the focus of US and European efforts against Russia (Bronson, 2014).

How this will be achieved remains unclear. Early discussions on exporting excess US liquefied natural gas (LNG) to Europe have been hampered by the lack of liquefied natural gas export facilities in the US for immediate distribution, lack of terminals in target countries and market pressures that drive exporters to Asia, not Europe. The facts have made it clear that increasing energy independence gives the United States new options, although a strategy to guide those choices is insufficiently underdeveloped (Griffiths, 2017).

Prior to the conflict in Eastern Europe and the world's focus on Russia's dominant energy position in Europe, the geopolitical effects of the rising US supply seemed to fall on traditional US partners, namely the Middle East. Questions were raised about the impact it may have on political relations with its Gulf partners. The declining relations between the US and Saudi Arabia are unexpectedly cited as a positive by-product of the new form of the US. Increased independence could thus have the positive effect of lowering US engagement and interest in the broader Middle East, and in particular for the Persian Gulf (Bronson, 2014).

But assuming the Persian Gulf partners would be less important to the United States because of their growing domestic supplies would be a bad choice. The lack of enthusiasm for rising US supply is a similarly serious analysis of rising global demand, which will have at least such a strong impact on the global energy market and shape the environment in which US policies are set. When it comes to both supply and demand, the Middle East emerges as a central element of US's ability to achieve its national interests. Washington will need its traditional partners to ensure safe energy supplies at reasonable prices or, in the current military language, safe trade flows. The debate about loosening ties between the United States and energy-producing Middle East partners as a result of US's growing energy independence is just that, loose and possibly reckless (Holland et al., 2015). All energy sources will be needed to weaken new demand, maintain prices further and promote global economic growth (Bronson, 2014).

There is no doubt that a revolution in the energy markets is taking place. Changing consumption patterns and developments in energy mining are profoundly changing US's global position. New shale gas and associated tight oil along with efficiency gains have helped the United States emerge as the world's largest producer of energy and reduce its dependence on foreign oil resources. By 2016, the United States is likely to import only 25% of its oil needs, up from 60% in 2005, though the rate will again slightly rise to over 30% by 2040. The net share of US imports of energy, gas and coal is projected to drop from 30% in 2012 to just 4% in 2040. This is due to increased domestic production and increased efficiency. Americans use less oil now than they did in 1973. Electricity use increased by just 6% between 2000 and 2012 and Americans used less total energy in 2012 than in 1999 despite a 25% higher economy in real terms. As a result, the United States is likely to approach energy independence before the end of the decade (Bronson, 2014).

Other countries such as Mexico, Canada and Brazil are expected to bring new and significant energy supply connections. In North America, Canadian oil will continue to enter the market with increasing volume. Mexico is drafting domestic legislation to take advantage of the technological developments of multinational corporations by reversing the decline in oil production and exploiting the potential of gas (Hinnebusch, 2015).

Brazil will be another major contributor to the new energy supply, driving South America's efforts in energy production. In the coming years, it will be ready to exploit huge reserves of almost 20,000 feet below the surface of the ocean. Across the globe, high energy prices and growing concern about the impacts of greenhouse gas emissions are causing technological change and encouraging new discoveries that will reshape the energy landscape as we know it (Bronson, 2014).

But the rising supply is fueled by an uncertain and growing energy demand. The relationship between the two will remain close for the foreseeable future. According to the International Energy Outlook of the Energy Information Administration (EIA), more than 85% of global energy demand growth from 2010 to 2040 will occur in countries outside the OECD, mainly in China and India (Mattis, 2018).

The slowdown in China's economic growth is unlikely to dramatically affect this trend. According to the IMF, growth in China increased in the second half of 2013, though this is likely to slow. But it will continue to rely on increased energy imports, despite China's huge investment in renewable energy sources. The EIA study shows that China will surpass the United States as the largest importer of oil (Bronson, 2014).

Growing demand from India will also put new pressure on energy markets. According to BP's Energy Outlook 2035, India's demand growth by 132% surpassed other countries such as Russia (+20%), China (+71%) and Brazil (+71%). India's growth is expected to almost double the total of non-OECD by 69%. To support this demand, BP expects that India's oil imports will increase by 169% and represent more than 60% of net increase in imports, followed by rising gas (+573%) and carbon (+85%) imports (Mattis, 2018).

Similarly, the EIA predicts that non-OECD consumption in Asia will increase from 148 to 337.5 Btu between 2009 and 2040, or an average of 2.5% per year over the same period. These numbers indicate a world that is seeking increased energy supply. According to the International Energy Agency, the Middle East remains the key factor in meeting this demand (Bronson, 2014).

1.2.3. The Geopolitics of Energy

International relations in the Middle East have long been dominated by uncertainty and conflict. External intervention, transnational wars, political unrest and inter-ethnic violence are exacerbated by the whims of oil prices and claims by military, nationalist and religious movements (Halliday, 2005).

Post-war Middle East literature usually treats regional conflicts and energy crises as related though substantially discrete phenomena. Wars are usually seen as the result of a combination of local conflicts complicated by superpower interactions. Energy crises, by contrast, are generally seen as a consequence of changing world market conditions and institutional arrangements. Some conflicts, such as the 1990-1991 war between Iraq and the US-led coalition, are partly due to a struggle to control energy reserves, while others, such as the 1967 and 1973 Arab-Israeli wars, and the Iraq-Iran conflict in 1980 were seen as exacerbating the ongoing energy crises (Fawcett, 2016).

So far, however, there has not been a single theory of energy conflicts, that is, a framework that integrates militarization and conflict with global energy flows and changing oil prices. In addition, existing literature on oil and war in the Middle East tends to address the potential role of fundraising rather poorly or not at all (Bidder & Nitzan, 1996).

From a static point of view, the 1990s seems to have brought the United States closer than ever to securing its national interests in the Middle East. After the Gulf War of 1990-1991, the United States began peace to promote reconciliation between Israel and its Arab neighbors. Since 1995, Israel has signed peace agreements with the PLO and Jordan, established diplomatic relations with Morocco and warm - albeit informal - relations with some of the Gulf Emirates, and negotiated a crucial agreement with Syria. The Gulf states had ratified US protection and access to their oil at low prices appeared to be secured (Bichler & Nintzan, 2014).

With the collapse of the Soviet Union and the rapid capitalization of emerging economies in Asia, Latin America and Eastern Europe, governments in the Middle East felt compelled to open their own economies to trade and investment. Soviet aid and military support have been doing their best, and the attention of industrialized countries is increasingly diverted to countries such as China, India and Brazil. In line with this new world order, many Arab elites have realized that they must integrate into the world market or risk the specter of economic crises and internal turmoil. Israel was in a similar situation, though its situation is much more comfortable now. Until the mid-1980s, the Israeli economy was characterized by a tight oligopolistic structure in which differential accumulation by the largest groups was supported through a regime of high military spending and rapid inflation. By the late 1980s, however, it became increasingly apparent that this economic order was no longer viable (Bidder & Nitzan, 1996).

Geopolitics is the struggle for space and power in a geographical environment. Just as there is military geopolitics, diplomatic geopolitics and economic geopolitics, there is also energy geopolitics. For the natural resources and trade routes that bring these resources to consumers, it is central to the study of geography. Every international order in early modern and modern history is based on an energy resource. While the Coal and Steam Era were the setting for the British Empire in the 18th and 19th centuries, the Oil Era was the setting for the American Empire from the late 19th to the early 21st century. And indeed, just as other countries and the American elites themselves were sending the United States into a period of decline, news began to emerge for huge shale discoveries in a number of states, especially in Texas. The Gas Era could make the United States the world's leading geopolitical power in the new century (Kaplan, 2014).

Mohan Malik, a professor at the Asian-Pacific Center for Security Studies in Honolulu, has been studying geopolitics for years. He has conceptually drawn up a new world map dominated by a growing consumer energy market in Asia and a growing production market in the United States. His research shows that in the next 20 years, 85% of the increase in energy consumption will come from the Indo-Pacific region. Already, at least a quarter of the world's liquid hydrocarbons are consumed by China, India, Japan and South Korea (Pascual, 2015).

According to the World Energy Report, published by the International Energy Agency, China will account for 40% of growing consumption by 2025, after which India will become the single largest source of growing demand. The growth rate of energy consumption for India will increase to

132%. In China and Brazil demand will rise by 71% and in Russia by 21%. The increase in gas demand will outstrip the increase in oil and coal. This means that the Indo-Pacific region will increasingly depend on the Middle East for its oil. By 2030, 80% of China's oil will come from the Middle East and 90% in the case of India. Japan and South Korea are still 100% dependent on oil imports. China's dependence on the Middle East will be offset by its subsequent and increasing dependence on the former Soviet Central Asian for energy (Kaplan, 2014).

While the Indo-Pacific region is becoming increasingly dependent on energy in the Middle East, in the other hemisphere, the United States is emerging as a global energy-producing giant on its own. Shale oil production will more than double between 2010 and 2020. The United States will open the Atlantic and Pacific coasts for drilling, and oil production in the United States and Canada could eventually be equivalent to both countries' consumption. Already within a decade, shale gas has grown from 2% to 37% of US gas production. The United States has now surpassed Russia as the world's largest gas producer. Some estimates show that the United States will surpass Saudi Arabia as the world's largest oil producer by the end of the current decade, though this is unlikely (Kim & Blank, 2015).

This would mean a return to the Yom Kippur war period before 1973, to US energy domination. When combined with Canadian and Brazilian oil, these shifts have the potential to make America the 21st century's new Middle East, though we must remember that US oil production may decline after 2020 (Kaplan, 2014) .

At the same time, Russia is increasingly shifting energy exports to East Asia. China is well on its way to becoming the largest oil export market in Russia before the end of the decade, even as Russian energy companies are developing closer ties with Japan in order to resist the growing emphasis they place on China (Bidder & Nitzan, 1996).

The Middle East will export more and more hydrocarbons to the Indo-Pacific region. Russia will export more and more hydrocarbons to the East Asian region, and North America will soon be increasingly considering the

Indo-Pacific region to export its own energy, especially gas (Kaplan, 2014).

As Indo-Pacific waters, namely the Great Indian Ocean and the South China Sea, become the world's energy network, tensions are rising in the South China Sea and the adjacent East China Sea. The territorial tensions for which the country holds the geographical characteristic of these waters are driven not only by the potential energy and fish stocks in the region, but also by the fact that the sea lanes are of increasing geopolitical importance due to the changing global energy market (Kaplan, 2014).

Europe, due to its aging population, is unlikely to increase its importance in the global energy markets, while the Indo-Pacific region will. Although northeast Asia, like Europe, is host to an aging population, this is not - or at least is less - the case in the Indian Ocean world (Scholten & Bosman, 2016).

Economic importance often leads to cultural and political importance over time. So today's tension between an economically and demographically stagnant European Union and a turbulent, authoritarian, energy-rich Russia may lead to the decline of Greater Europe, while North America and the Indian Ocean make the world the new vibrant trade center. At the same time, however, at least in the short term, there appears an alliance between Russia and China under the auspices of a growing energy relationship as these two huge Eurasian states clash and compete with the democratic West (Scholten & Bosman, 2016).

Therefore, power in Eurasia will shift more to southern latitudes, while the United States will revitalize its own power with an even closer economic relationship with Canada and Mexico, which are also energy-rich countries (Kaplan, 2014).

1.3. The European Energy Security

According to the International Energy Agency (IEA), energy security is "the uninterrupted availability of energy sources at an affordable price." Also, in order to better fit the current political environment going through the threat of climate change and terrorism, the concept of reliability and sustainability was added to the definition. There is not only long-term energy security, which mainly deals with timely investments to supply energy in line with economic developments and environmental needs, but also short-term energy security focuses on the ability of the energy system to react promptly to sudden changes in the supply-demand balance. Energy security can be either used as a goal or as an instrument to shape foreign policy. Foreign policy defines the policy pursued by a nation in its dealings with other nations, designed to achieve national objectives. The European Union energy policies are an essential part of the European Union's strategy. They are meant to secure energy supplies, ensure there is competition for affordable prices, and to make sure the EU's energy consumption is sustainable. This, way the EU aims at being less dependent on its energy imports. As a matter of fact, the EU is currently importing more than the half of its energy (54%) costing more than €1 billion per day. Energy represents 20% of the total EU imports. The EU imports 90% of its crude oil and 69% of its natural gas. The EU imports a third of its natural gas from Russia. Germany, Italy, France, Belgium, and Spain are the countries importing the most their natural gas from Russia, Norway, Algeria and Qatar. However, Norway's natural gas supply is declining and Algeria's natural gas future exports to the EU are uncertain as their natural gas contracts will come to an end in 2019 and 2020. Qatar will most likely remain an important natural gas supplier for the EU countries in its capacity to sell LNG. In fact, the EU countries that are the most dependent on Russian gas owe it to their own domestic production, the share of natural gas in their energy mix, geographic location, political relationships, and the possibility for them to find alternative supply countries. Estonia, Finland, Latvia, and Lithuania, importing all their natural gas supply from Russia, are thus more exposed to energy supply disruptions. Despite the fact that one of the EU's main purpose is to become less dependent on Russia, according to Gazprom's delivery statistics, the Russian energy company has delivered its highest natural quantity to the European Countries in 2017, representing 192.2 billion cubic meters of natural gas. 81% of Gazprom's 2017 natural gas was delivered to the Western

European countries while 19% was delivered to the central European states. According to Valery Nesterov, an oil and gas analyst at Russian bank Sberbank, this rise in the EU demand for gas is due to the European economic recovery, gas prices being more competitive than those of other energy resources (notably the price of coal), the cold winter, the decrease of the Dutch natural gas output, and the ending of nuclear energy production in Germany. The EU is seeking to diversify its energy sources to avoid issues such as the series of gas crisis between Moscow and Kiev that has impacted negatively Russia's natural gas deliveries to Europe. As a matter of fact, in 2014 Gazprom cut off its exports to Ukraine. This resonated as a call for the EU to secure its energy supply through the European Energy Security Strategy. Nowadays, tensions with Russia being at their highest since the Cold War, the EU has good reason to fear that Russia uses its natural gas leverage as political means. Therefore, the EU Energy Security Strategy first consists of meliorating the cooperation of the EU countries in terms of natural gas distribution. If a natural gas disruption from Russia was to come, the cooperation between the EU countries could cope with it and be able to provide energy to its consumers for around six months. Also, part of the EU Energy Security Strategy is to increase energy efficiency, increase the energy production in the EU and diversify suppliers and routes, completing the internal energy market and building missing infrastructures, coordinating the member states' external energy policy, and strengthening emergency and solidarity mechanisms and protecting critical infrastructures, ("Energy Security Strategy," 2014). The EU countries also agreed on a 2030 Energy Strategy framework for climate and energy. This framework is made of targets and objectives in order to achieve a more competitive, secure and sustainable energy system for the period of 2020 to 2030. This strategy shows that the EU aims at attracting private investments in new pipelines, electricity networks, and low carbon technologies. The targets for 2030 are to reach a 40% cut in greenhouse gases emissions compared to 1990 levels, at least a 27% share of the renewable energy consumption, and at least a 27% energy savings compared with the business-as-usual scenario. In order to reach this purpose, the European Commission proposes a reformed EU emissions trading scheme, new indicators for the competitiveness and security of the energy system, as well as a new governance system following a common EU approach and based on national plans for competitive, secure, and sustainable energy ("Israel, Cyprus, and Greece push East Med gas pipeline to Europe," 2018). The EU shows more and more its interest for the eastern Mediterranean natural gas. The EU expressed its will to assist the countries in the region in the exploitation of their energy resources, as well as to develop commercial cooperation with those countries. As an example, the EU commission developed the CyprusGas2EU project due to begin on June 2018. This project consists of integrating the Republic of Cyprus at the core of the European Union energy strategy by installing the required infrastructure to send the eastern Mediterranean natural gas to its member states. The project's investments include establishing "a floating storage and regasification unit (FSRU), a jetty for the unit's safe mooring, a shelter for the FSRU and LNG carrier, a jetty borne pipeline and an onshore gas pipeline, an onshore storage array and a metering station at the Limassol port, terminal 2 (Vassilikos)," (2018). However, this EU project to invest more in the Mediterranean natural gas is not to the taste of environmentalist groups. The EU therefore seems to balance it with a communication strategy highlighting the potential growth and creation of jobs these projects would bring. Also, the EU commission's press release on the project took some liberties with the definition of clean energy: "More growth and jobs: EU invests €873 million in clean energy infrastructure." As a matter of fact, natural gas remains in the category of fossil fuels and releases greenhouse gases, it thus cannot be considered as a clean energy. Through the exploitation and use of the eastern Mediterranean discoveries, the EU has the opportunity to promote its internal cohesion, and its energy integration, which are especially important for the southern and eastern EU member states. Nonetheless, escaping the dependence from the Russian gas may be more complicated than it seems. Indeed, Russia's natural gas is the cheapest. As Russia also hold interests in keeping the EU countries as its importers, it may even follow the competition and further reduce its prices. The EU will probably have to pick a side between cheaper natural gas and its energy security. It is the EU's regulatory frameworks and policies that will shaping the competitiveness of the natural gas market that will later determine whether the EU will succeed to meet its energy security objectives. In order to facilitate the natural gas market development in the eastern Mediterranean, the EU could take the role of an ambassador to help building regional cooperation. As a secure and reliable actor in the natural gas market, the EU can facilitate and set incentives for the other actors in the region to better negotiate and come to necessary agreements. On the economic side as well, the EU could help the region stakeholders to attract investors. The EU is already engaging multilateral discussions through the Union for the Mediterranean (UfM). However, this institution is yet solely used as a discussion platform, and it may be interesting to further develop its concept and build on it to take active measures towards regional cooperation on energy.

Chapter 2

ENERGY, ENERGY RESOURCES, ENERGY SECURITY AND ITS BASIC PRINCIPLES

'Nothing exists that is not energy,
or not affected by energy.'
(Ciuta 2010, 124)

2.1. INTRODUCTION

This chapter presents considerations on energy security, its importance, the disruption in energy supply, and finally on energy crises. Furthermore, in the beginning, it clarifies concepts such as energy, energy resources which can be classified as non-renewable and renewable ones.

2.2. ENERGY, ENERGY RESOURCES

Energy is a good that became a need. There cannot be evolution or dignified living conditions without it. For this reason, the need of all countries to have access to energy is vital. Producing countries are not so many and their connection to all countries to transfer energy is protected by law - the Energy Charter Treaty and International laws for energy -, by contracts and agreements. When a state agrees with other to pass the energy through its land or water territory by pipelines or ships, in order to transit to another country, it becomes a transit state.

There are two key factors which play crucial role for a state to have the right to become a transit state. The first one is, evidently, its geographical data. In other words, it depends on the position and the boundaries that a place has. It should be near to a producer country or another transit country so as to pass the energy from them to its territory and then to others. As a result, countries, which are at a central - nodal area, have more possibilities to play such a role. The second, maybe even more defining key is the diplomacy of the country and the international relations with others. The stability and the sound conditions between the states boost the aim to reach an agreement of high guarantees. Relevant to the latter and of equal seriousness are the wealth, the sources of energy and the state of the economy, the strength of the Armed Force, the size of the population, the participation in international organizations and international financial associations (NATO, EU, OECD, etc.), the polity and social serenity, the possibility to pressure foreign governments successfully, the investments made in foreign countries, the culture and cultural influence abroad, the overall international awards academic and other conquests and the infrastructure of the new technologies.

One fundamental principle in the sector of energy is the unbundling - vertical or horizontal integration - after the liberalization of the market, which means separated providers for the various procedures. This happened to break the monopolistic regime and Production, Transmission, Distribution, Trading, storage; supply cannot be elaborated by the same "player." More specifically, it is necessary that at least a few states transfer energy and secure access to the commodity for others. Everyone needs to have access to the grid and transit and hub countries keep this role. Diversification, stability, and protection are some of the most important issues combined with Energy Security.

The Ukrainic crisis is a characteristic example of how important the existence of diverse ways is retaining and securing energy supply. Due to their limited energy sources, EU countries have a high rate of dependence on external sources, mainly from Russia. Faced with this situation, EU countries have taken several steps to diversify energy sources as a major element in ensuring energy security. Developing integrated networks of gas and power supply throughout Europe, including the interconnectivity of pipelines and the increase of oil and natural gas storage capacity are encouraging signs in this direction.

Before proceeding to an analysis of energy security and its principles, let's look at the most important non-renewable and renewable sources of energy.

Non-renewable energy resources are resources that are not replenished or that will not be replenished for thousands or even millions of years. Non-renewable energy sources include coal, oil and natural gas, also known as fossil fuels. For non-renewable energy sources we refer to oil as it is the main source of energy used in many forms and the raw material for many products, to natural gas that has the advantage of being transported over long distances through pipelines, and to hydrates actually methane hydrates that are one of the most promising energy sources for the 21st century.

Renewable energy sources also called renewables, are energy sources that replenish (or renew) themselves naturally. Typical examples are solar energy, wind energy, hydroelectric energy and biomass.

An energy source that has been little used to date is the energy of the sea. The sea surface absorbs huge amounts of solar and wind energy, that appear in various forms, such as waves or currents. The wave energy – inexhaustible and of high density –, the impressive energy, based on the use of semi-permeable films, the thermal energy, that depends on the temperature difference between the warmest surface water and the coldest bottom water, and the wind power, they all belong to renewable energy sources. (Arena et al., 2015; Bagiorgas et al 2012).

2.3. Types of Energy

Before proceeding to an analysis of energy security and its principles, let's take a look at the most important non-renewable and renewable sources of energy.

2.3.1. Non-Renewable Energy Resources

Non-renewable energy resources are resources that are not replenished or that will not be replenished for thousands or even millions of years. Non-renewable energy sources include coal, oil and natural gas, also known as fossil fuels.

2.3.1.1. Coal

The term "coal" refers to organic sediments derived from plant residues through a series of carbonation processes. These processes have resulted in the enrichment of plant residues in carbon. The conversion of plants into peat and the transition from peat (initial carbonation stage) to anthracite (final carbonation stage) is a function of the effect of time, temperature and pressure. The conversion of plant material into carbon began about 400 million years ago and continues to this day. Experts estimate that a 2.5-meter layer of plant material is needed to create a 30-centimeter layer of carbon.

The classification of coal is determined by its calorific value in combination with the chemical analysis of organic matter. Carbohydrates with high carbon content and low hydrogen and oxygen content are characterized as high quality while the reduction in carbon content reduces the quality of the coal. Depending on the degree of carbonation, the coals are divided into peat, lignite, underlying coal, rear coal and anthracite. The transition from peat to anthracite increases the thermogenic capacity of the carbon and of course their quality as a source of energy (Ayat, 2013).

Almost every country on the planet has coal reserves, but only in 70 countries worldwide is coal mining a commercial activity. Given current levels of production and consumption, global coal reserves are sufficient for the next 164 years. In contrast, oil and gas reserves are sufficient for the next 41 and 67 years, respectively. In the medium term, a significant increase in carbon consumption is expected in developing countries, especially in China and India. By 2030, these two countries are estimated to be responsible for the consumption of almost two-thirds of the global coal reserves.

2.3.1.2. Oil

Oil is found in the subsoil in liquid form, in cavities, formed there by animal and plant microorganisms, mainly marine ones, which were collected by sea currents in the depths of basins, where they were crushed due to earthworks or other processes. There, without the presence of air, they were converted to oil over millions of years.

The energy of oil comes from the energy that the microorganisms that created it had gathered from the sun and their food. Today, oil is extracted from its underground deposits, even if they are under the seabed. Its main components are alkanes (paraffins), cyclohexane (naphthenia) and aromatic hydrocarbons and in smaller quantities oxygen, nitrogen and sulfur compounds.

Oil is the most important mineral for the global economy, as it is the main source of energy and the raw material from which a huge number of products are produced [plastics, medicines, cosmetics, detergents, film tapes, explosives, etc.) (Jensen et. al., 20120)].

2.3.1.3. Natural Gas

Natural gas is a mixture of hydrocarbons and consists mainly of methane and a much smaller proportion of ethane, propane, butane and pentane. A determining factor in its composition is its origin, especially if it is a purely natural gas field or is derived from oil fields.

Its commercial use began around 1810 as a fuel for light bulbs, and after the end of World War II, the first gas transmission and distribution networks were built. Its advantages as an energy source include its ability to be transported over long distances through pipelines and of course its comparatively environmentally friendly combustion (Shaffer, 2011).

2.3.1.4. Gas Hydrides

In recent years, a type of natural gas that has started to play a key role in the global energy market is the so-call shale gas. The term shale gas refers to natural gas that is trapped within shale formations.

Oil and gas researchers are working to develop new technologies to produce natural gas from methane hydrate deposits.

After the energy revolution of shale gas, another form of non-conventional natural gas, methane hydrates, follows. Methane hydrates are one of the emerging clean energy sources of the 21st century. Methane hydrates grow in areas with a permanent layer of ice on the ground or below the seabed. About 98% of hydrides are found in the substrates of the seabed. These are usually covered by a layer of sediment. Their formations below the seabed require an environment with fairly high pressure (pressures greater than 30 bar or 30 times the normal atmospheric pressure) and low temperature (temperatures below 10 degrees Celsius). The hotter the water, the higher the water pressure. Thus, in the Arctic, methane hydrates can be found below water depths ranging from 200 to 3,000 meters, while in the tropics they can be found only below 600 meters. Methane is surrounded by water molecules and trapped in a molecular cage. Therefore, Chemists call this type of molecular structure clarithromycin (Urgeles & Camerlenghi, 2013).

When they come out of the bottom of the sea, they start to break slowly. This releases methane gas which can then ignite.

Three different processes are used for methane recovery (Urgeles & Camerlenghi, 2013). The first one is using water circulation. Hot water is pumped into the hydrogen methane field through a well, raising the temperature to the point where the hydrates decompose and methane is released. The second one is based on decompression. Given the fact that high pressures predominate in the wet methane layers, due to the excessive water and sediment loads, drilling is done in the deposits from above, releasing the pressure in a way similar to the drilling of the pipe into a bicycle tire. As the pressure drops, the hydrates break down slowly and the methane is released. The last one makes use of carbon dioxide injections. Methane is released by hydrates when impregnated with a gas. Carbon dioxide displaces methane in the inclusion compound, replacing it in the molecular cage. A result of this is a stronger beam of the water molecule with carbon dioxide than it had with methane. It turns out that the carbon dioxide hydrate is significantly more stable than the methane hydrate.

Although only rough estimates can be made about global reserves of this new form of non-renewable energy, since it is new, experts opinions concur that our planet contains large amounts of methane hydrates. The potential water reserves are estimated at 21,000 in trillion cubic meters, which is about 100 times the total gas reserves in the world so far. Assuming that only 10% of all reserves is used, it can last for about 600 years.

The presence of hydrates in the European continent is found so far in the Caspian Sea, in the Barents Straits and in the Eastern Mediterranean, in the submarine shells of Anaximander, east of Rhodes and south of the Gulf of Antalya and Kastelorizo.

Some argue that the existence of methane hydrates may be the real reason for the collision of all geopolitical actors in the region of the Southeastern Mediterranean and not the conventional gas and oil reserves.

Possible deposits of methane hydrates have been reported in marine sediments in Nankai Trough off the Pacific coast in central Japan, where the water depth is over 500 meters. Some estimates suggest that methane hydrate reserves correspond to a 100-year supply of natural gas for Japan, making it an important potential source of energy (Lu, 2015).

It is possible that after 2020, methane hydrate mining technology will become commercially viable and the cost of mining sustainable. The above will mean significant geopolitical changes including: a) The general detoxification of Japan from gas imports and perhaps the partial detachment of China, b) the gradual marginalization of conventional hydrocarbon production countries such as Saudi Arabia and Qatar, but not of Russia, which has several possible areas for methane hydrate deposits, c) the future supply of natural gas to the EU from the production of methane hydrates coming from the Black Sea and the Southeastern Mediterranean, and last but not least, d) an increase in the geopolitical importance ofSoutheast Mediterranean, Southeast Asia, Black Sea, Caspian Sea, North Sea and of course the seas around India that seem to contain large amounts of methane hydrates. (Urgeles & Camerlenghi, 2013).

2.3.2. Renewable Energy Resources

Renewable energy sources, also called renewables, are energy sources that replenish (or renew) themselves naturally. Typical examples are solar energy, wind energy, hydroelectric energy and biomass.

2.3.2.1. Solar Energy

Solar energy is energy from the sun that is converted into thermal or electrical energy. Solar energy is the cleanest and most abundant renewable energy source available. Solar technologies can exploit this energy for a variety of uses, including generating electricity, providing light and heating water for domestic, commercial, or industrial use.

There are three main ways to harness solar energy: photovoltaics, solar heating & cooling, and concentration of solar power. Photovoltaics generate electricity directly from sunlight via an electronic process and can be used to power anything from calculators and road signs to homes and businesses. Solar heating & cooling (SHC) and concentrating solar power (CSP) applications both use the heat generated by the sun to provide space or water heating in the case of SHC systems, or to run traditional electricity-generating turbines in the case of CSP power plants.

2.3.2.2. Wind Energy

In recent years, there has been a spectacular rise in installed electricity from wind turbines in our country (Crete, Evia, island country). The rapid growth of wind energy was accompanied, as it was next, by the concern of local communities about the possible effects of wind turbines on the environment.

In some cases, the fears expressed are rather exaggerated and sometimes unrealistic. In other cases, the objections to the installation of wind turbines or wind farms have some basis and need further investigation. In any case, the acceptance or non-acceptance of wind energy by local communities presupposes their objective information about the benefits and effects that this could have as another human intervention in nature (Furevik et al., 2011).

As with all RES, wind energy is a clean and environmentally friendly form of energy. Its exploitation using wind farms does not pollute the ecosystems of the installation areas. Land requirements for the construction of wind farms may be high, but during operation they take up little space.

Thus, the land remains available for other uses, while the passage of the fauna is free. The technology of wind turbines is now going through the first period of "maturity" which makes it economically competitive in terms of initial investment and operating costs. At the same time, their assembly and installation is fast and standard (Furevik et al., 2011).

Placing them in more remote areas with rich wind potential (such as the Aegean islands) helps to decentralize the energy system by reducing energy transfer losses from the central distribution network and increasing the independence of the regions from it. Especially in Greece, the significant dispersion and inhomogeneity of electricity generation costs in different parts of the country, results that even if the average price of electricity is slightly lower than the marginal cost of wind produced kWh, in several island areas , the production cost of electricity is multiple, sometimes more than tenfold, of the marginal production cost of PPC.

But even if the cost of generating electricity from wind is about twice the cost of electricity generated by lignite, its environmental impact is incomparably smaller than the production of electricity from fuel. Wind farms also contribute to the energy independence of each country from imported fuels, which in turn lead to foreign exchange hemorrhage, on the one hand, and dependence on other states on the other.

The estimated EPA can reach and exceed with the gradual acquisition of experience and 90% of the total cost of a wind turbine, while strengthening the domestic economy. At the same time, they contribute to reducing unemployment by creating new jobs in the region (Bagiorgas et al., 2012).

Adversaries of wind technology report a number of issues that need to be taken into account. So the first issue is the noise emitted by wind turbines. Depending on its origin, it can be divided into two categories: a) Mechanically - coming from the rotating mechanical parts (gearbox, generator, bearings, etc.), b) Aerodynamically - from the rotation of the fins.

In terms of environmental awareness it has also been stated that this acoustic disturbance can have detrimental effects on the fauna, disturbing the balance in the perpetrator-prey system and interfering with the communication sounds emitted. orientation.

2.3.2.3. Hydroelectric Energy

Hydroelectric energy, also known as hydroelectric power or hydroelectricity, is a form of energy that exploits the power of water in motion (such as the water of a waterfall) to generate electricity. People have used this force for millennia. More than two thousand years ago, people in Greece used flowing water to turn the wheel of their mill to ground wheat into flour.

The way hydroelectric energy works is this. Most hydroelectric power plants have a reservoir of water a gate or valve to control how much water flows out of the reservoir, and an outlet or place where the water ends up after flowing downward. Water gains potential energy just before it spills over the top of a dam or flows down a hill. The potential energy is converted into kinetic energy as water flows downhill. The water can be used to turn the blades of a turbine to generate electricity, which is distributed to the power plant's customers.

Hydroelectric energy is the most commonly-used renewable source of electricity. China is the largest producer of hydroelectricity. Other top producers of hydropower around the world include the United States,

Brazil, Canada, India, and Russia. Approximately 71 percent of all of the renewable electricity generated on Earth is from hydropower.

2.3.2.4. Geothermal Energy

The term geothermal energy is used to refer to the thermal energy that comes from inside the earth. Geothermal energy is defined as the exploitation of energy from the interior of the earth from which the use of a geothermal heat pump allows the transfer of heat from and to the ground to produce cooling, heating and hot water for domestic and wider applications. . The use of the region's volcanoes, the so-called volcanic arc in the Eastern Mediterranean, could provide energy dynamics through geothermal energy.

2.3.2.5. Bioenergy

Bioenergy is a form of renewable energy derived from the utilization of recently living organic materials of biological origin known as biomass, which can be used to produce transportation fuels, heat, electricity, and products. The term Biomass is used to refer to raw materials that, through the process of photosynthesis, have converted the solar energy they receive through radiation, into chemical. By its definition, biomass includes a large number of different materials, such as energy plants (sage, olive, corn, hemp, sorghum, wild artichoke, etc.), wood in all its forms (waste wood, sawdust, pruning, etc.), agricultural by-products (branches, straw, kernel wood, etc.), livestock waste, organic industrial waste (whey, slaughterhouse waste, goat milk, etc.), food waste (food waste, by-products of industrial processes, fats, oils, etc.) (Diaconu & Rusu, 2013).

The fuel produced from the proper processing of biomass is called biofuel. In essence, biomass is the raw material from which biofuel is made of and biofuels can be burned directly to generate electricity, and thus bioenergy. Therefore, the separation of biofuels from bioenergy is meaningless as we are essentially referring to the same thing, that is tothe sustainable production of biomass energy. It should be noted that second-generation biofuels, which are increasingly used to produce bioenergy, do not have a negative effect on the food chain, as they use waste or non-edible plants to produce fuel.

2.3.2.6. Marine Energy

One renewable energy source that has been little used to date is marine energy. Marine masses cover 75% of the planet's surface. The sea surface absorbs huge amounts of solar and wind energy, which appears in the sea in various forms, such as waves or currents. In addition, the marine system is affected by the planet's gravitational interactions with the Sun and the Moon.

Marine energy or marine power (also sometimes referred to as ocean energy or ocean power) refers to the energy carried by ocean waves, tides, salinity, and ocean temperature differences. The movement of water in the world's oceans creates a vast store of kinetic energy, or energy in motion. This energy can be exploited to generate electricity to power homes, transport and industries.

There are many forms of marine energy and huge amounts of energy that can be used. A common property of marine energy forms is the high energy density, which is the highest among renewables. Today, various technologies have reached such a stage of technical maturation that the mass use of the sea for the production of "pure" and "cheap" energy is now considered feasible (Shields et al., 2011).

2.3.2.7. Wave Energy

The energy of sea waves is inexhaustible. It is estimated that using 1% of our planet's wave potential would quadruple global energy demand. It has the highest energy density among renewables.

Among the various forms of ripple, wind turbidity is of the greatest interest for energy exploitation. Windswept waves are created by the interaction of the wind with the sea surface. If wind turbulence is created, it can "travel" thousands of kilometers with minimal losses (Boudière et al., 2013). The sea waves are caused by the air as it blows beyond the sea. Waves are a powerful source of energy. The problem is that it is not easy to use this energy to convert large amounts of electricity. as a result, power generation stations are rare.

Although systematic research into the exploitation of wave energy has been underway for decades, the relevant technologies have not yet reached the stage of commercial exploitation. The main reason is the unfavorable environment, which is disruptive and has slowed growth in this area. However, the efforts of the past decades have begun to bear fruit and a number of the technologieshave now reached such a stage of technical "maturation" that in the short term their mass installation could begin for the electrification of coastal areas, islands, etc. (Arena et al., 2015).

2.3.2.8. Impressive Energy

In theory, the application of the osmosis principle for energy production was first developed by Israeli researcher Sidney Loeb in collaboration with Srinivasa Sourirajan in the 1970s. The key to its realization is the semi-permeable film, which must be slightly permeable. from the side of fresh water, while from that of salty to maintain the growing pressure steadily on its own side. Increasing the pressure on the salt water side can be used to operate a turbine to generate electricity.

The first implementation of the osmosis phenomenon for large-scale electricity generation took place in 2009 by the Norwegian company Statkraft, which operates in the field of RES. The factory under test started operating on November 24, 2009 with a power of only 2 - 4kW. However, the company expects to produce 25MW of electricity by 2015 using osmotic energy. The company's executives claimed that osmotic energy could produce about 1600 - 1700TWh worldwide - the equivalent of half the energy produced in the EU. today (Takao & Setoguchi, 2012).

2.3.2.9. Ocean Thermal Energy

Ocean thermal energy can also be exploited by exploiting the temperature difference between the warmest surface water and the coldest bottom water. This difference must be at least 3.5. C to be exploited by a heat engine.

2.3.3. Nuclear Energy

Nuclear energy is the energy released when atomic nuclei are transformed. In other words, it is the dynamic energy that is trapped in the nuclei of atoms due to the interaction of the particles that make them up. Nuclear energy is released during fission or fusion of nuclei, and if nuclear reactions are controlled (as happens in the heart of a nuclear reactor) it can be used to meet energy needs. In most cases the fuel used for nuclear fission is uranium.

Although nuclear energy is considered clean energy, its inclusion in the renewable energy list is a subject of major debate, mainly due to the fact that uranium deposit on earth is finite and that the nuclear waste from nuclear power reactors is considered as a radioactive pollutant, and that goes against the notion of a renewable energy source.

2.3.4. Electricity Insurance

Electricity is defined as the flow of electric power or charge. It constitutes a basic part of nature and it is one of the most widely used forms of energy. We get electricity, which is a secondary source of energy, from the conversion of primary sources of energy, such as coal, natural gas, nuclear energy, solar energy, and wind energy, into electrical power. Electricity is also referred to as an energy carrier, meaning that it can be converted into other forms of energy such as mechanical energy or heat.

The modern world depends on this kind of energy for its survival and prosperity. Most devices are powered by electricity. There are many ways to generate electricity. The main ones include the combustion of various substances (lignite, oil, coal), nuclear power plants, solar parks, hydroelectric dams and wind farms. In the last 20 years, intensive efforts have been made to increase the percentage of electricity generated by the use of Renewable Energy Sources (RES).

The big disadvantage of electricity is that it cannot be stored for a long period of time.. For this reason it should be consumed soon after production or it must first be converted into other forms of energy (eg chemical, dynamic, etc.) in order for it to be stored The need for direct electricity consumption has led to the construction of a global energy grid through the interconnection of electric power networks between regions and continents, so that electricity can be easily transported from its point of production to the point of consumption (Takao & Setoguchi, 2012).

2.4. Security

The concept of "security" and the definition of security issues has been, and continue to be, one of the most ambiguous and difficult-to-define issues in International Relations. At the same time, the concept of security remains central and prominent in International Relations, adopted by all theoretical movements, serving as an analytical basis for issues on the internal political organization of state entities, on the international environment and on transnational relations.

Security is traditionally defined as the absence of threats. "Security, from an objective point of view, is defined as the absence of threats to acquired values and, sense from many people's subjective point of view, as the absence of the fear that those values are going to be affected." (Borchert & Forster, 2007).

The concept of security is a fundamental concept of both the realistic and the idealistic school of thought in international relations. It offers an alternative framework of analysis in which central concepts, such as war, peace and their causes, and the conclusions that lead to the two streams of thought can be reconciled.

Traditional security focuses on state and military power and its relation to the "national security" of the nation-state. Maintaining statehood, national independence and territorial integrity are central values that other states may threaten with the use of military means.

National security, also called "military security," refers to the states' defense capabilities and their intentions against threats from other states.

Military security and related issues and dilemmas are also called "conventional" security issues/dilemmas.

2.5. Identifications

Energy security, as an autonomous security sector, is a very modern concept, like that of environmental security. Energy security is a complex concept that evolved over the time. Several authors indicate to the absence of the clear definition of the term. Energy issues are intersecting in the areas of economic and environmental security and are of strategic importance for the national security of states. Following table shows schematically and succinctly how energy issues are an important aspect of the different definitions of energy security based on input from several researchers. The concept of energy security is an inherent term of many meanings, versions and dimensions depending on the geographical region (country, region, and continent), the time horizon and the particular energy source to which it refers. It is a highly politicized and multi-faceted concept (McCormick, 2013).

2.6. Energy Security Crisis

At this point it should be noted that the term energy security is not static. It is clearly governed by some basic principles, but new components are constantly being added, mainly due to international political conditions, technological progress and the shortage of energy resources. The concept of energy security is interrelated with perspective and largely depends on the context in which it is defined. The multitude of definitions of energy security allows each energy actor (states, businesses and ordinary consumers) to develop their own different interpretation of energy security and based on it, choose the ways and means to satisfy it (López et al., 2013).

Table 2.1. Security areas and Energy Sector Relationships (López et al., 2013)

Security areas	Main object	Energy Issues
Military	• Maintaining statehood and territorial integrity vis-à-vis military threats from third countries • Strengthening the state's defense capabilities	• Control of energy sources can be the main cause of military conflicts • Energy products necessary for the proper functioning of the defense equipment • Energy
Politics	• Preservation of national sovereignty • Maintain the organic stability and institutional legitimacy of the state	• High dependence on third country energy sources threatens national sovereignty
Economic	• Finding the necessary resources and capital to maintain the generally accepted levels of prosperity and the proper functioning of markets and state power	• Energy products are the basis of the economic system and the driving force of the industry • Disruptions in global energy markets (eg oil) can adversely affect national economies
Social	• Maintaining conditions that promote the progress and cohesion of national identity	• Energy poverty and unequal distribution of energy resources between different groups of the population can cause internal imbalances
Environmental	• Conservation of sustainable conditions of the local and global biosphere that is the domain of human activity development	• The energy mix is responsible for the environmental degradation levels • Transmission or Power Generation Accidents (Oil Spills) Causes of Environmental Pollution

Table 2.2. Basic principles of energy security (López et al., 2013)

Basic Principles	Structures	Possible Diversions
Availability	• Geological factor. Producer countries' natural energy reserves • Agreement between country producers, transit centers and final consumers on terms of trade in energy sources • Technology available for the production, transport, distribution, conversion and storage of energy sources • Capital investment in energy-related sectors • Existence of stable legal and regulatory structures • Compliance with environmental requirements	• Depletion of energy reserves that are exploitable and can be exported at a relatively low cost • General restrictions on the development of the energy sector due to national choices and transnational differences, in particular as regards transit • Economic, legal, regulatory, political structures in general, which do not favor sustainable energy investment
Reliability	• Diversification-based energy supply chain • Adequacy of stocks to cover the transit chain in its entirety • Short-term and long-term protection from terrorist attacks, extreme weather and political conflicts • Adequacy through the exchange of information on the operating conditions of energy markets	• Inability of transit systems to weather and other adverse environmental conditions (e.g., earthquakes) • Poor infrastructure maintenance or insufficient investment • Attack (or threat of attack) by terrorist organizations or other armed forces • Political interference (transnational conflicts, embargo or sanctions, etc.)

Basic Principles	Structures	Possible Diversions
Accessibility	• Existence of well-functioning energy markets in the context of which energy prices can be adjusted • Slight price volatility • Transparency in pricing processes and realistic expectations about future price fluctuations	• Depletion of energy reserves that can be exported at a relatively low cost • High energy demand and inability to adjust the pricing system
Environmental sustainability	• Low emissions of greenhouse gases and other pollutants to preserve the quality of the environment at local and regional level • Protecting energy systems from the effects of climate change	• Develop energy policies in the light of energy security without considering environmental factors • Negative impacts of climate change on energy infrastructure

2.7. Disorder of Supply

The planet is facing two energy crises. The first one comes from the limited energy sources and the traditional power policies applied, where the larger and more active economies struggle to acquire the natural resources they need (Shields et al., 2011). The second one is new; it relates to the climate change and it requires international cooperation rather than competition. Two conflicting crises, with no visible solution on the horizon. While the first crisis is pushing politicians and businessmen to seek more oil and gas, the second is calling for a drastic reduction of dependence on hydrocarbons.

Politicians are facing two crisis hotspots. Former British Prime Minister Tony Blair had been pushing for an international agreement on climate change for months. But he believes one of the most important - and less audible - achievements of his term was to secure a long-term deal to supply Britain with natural gas from Norway. Theoretically, the two energy crises point to the same direction. Developing alternative, "clean" forms of energy will reduce dependence on oil and gas. This is also the determining factor for any effort to reduce emissions of pollutant gases. The problem is that there is little evidence that alternative energy can be developed quickly enough to meet oil and gas demand (Boudière et al., 2013).

2.8. Law of the Sea, Continental Shelves and EEZ

The law of the Sea is a body of international law that states all the principles and rules of which public entities (especially states) interact in maritime matters, including navigational rights, sea mineral rights, and coastal waters jurisdiction. The law of the Sea displays the rules that regulate the entitlement of coastal states to maritime zones, their rights and duties within those zones and how the boundaries of each zone should be established. Much of its law is codified in the United Nations th

Convention on the Law of the Sea, signed on the 10 of December 1982. The convention is often described as the "Constitution for the oceans" and it is aiming at codifying international laws regarding territorial waters, sea lanes and ocean resources ("United Nations Convention on the Law of the Sea," 1982). This body of international law is important as it is setting a framework for the exploitation rights on the reservoirs of resources that can be found in the oceans and seas. The fundamental principle of it is that the land dominates the sea. This means that, it is the land's territorial situation that determines the maritime rights of a coastal state. The determination of the maritime rights is of utter strategic importance as the outer continental shelf hosts more than 50% of the world's remaining yet-to- be- discovered natural gas resources.

As legally stated in the UNCLOS, the exclusive economic zone is "an area beyond and adjacent to the territorial sea, subject to the specific legal regime established in this Part, under which the rights and jurisdiction of the coastal state and the rights and freedoms of other states are governed by the relevant provisions of this Convention.," ("United Nations Convention on the Law of the Sea," 1982). That is to say that an EEZ is a zone of coastal water and seabed distanced from 200 nautical miles (370 km) from the shore (baseline) of the country claiming the rights on this zone notably for commercial activities, fishing, drilling. Also, it is adjacent to the 12- nautical mile territorial sea (See Figure 2.1). A state benefiting from the EEZ owns the right to exploit and regulate fisheries, construct artificial islands and installations, use the zone for other economic purposes, and regulate scientific researches by foreign vessels. Without such regulations, foreign vessels and aircrafts have the right to move freely through the zone. Also, still defined by the UNCLOS, the continental shelf is constituted of the seabed and subsoil of the submarine areas that extend beyond its territorial sea throughout the natural prolongation of its land territory to the outer edge of the continental margin, or to a distance of 200 nautical miles from the baselines from which the breadth of the territorial sea is measured where the outer edge of the continental margin does not extend up to that distance, ("United Nations Convention on the Law of the Sea," 1982).

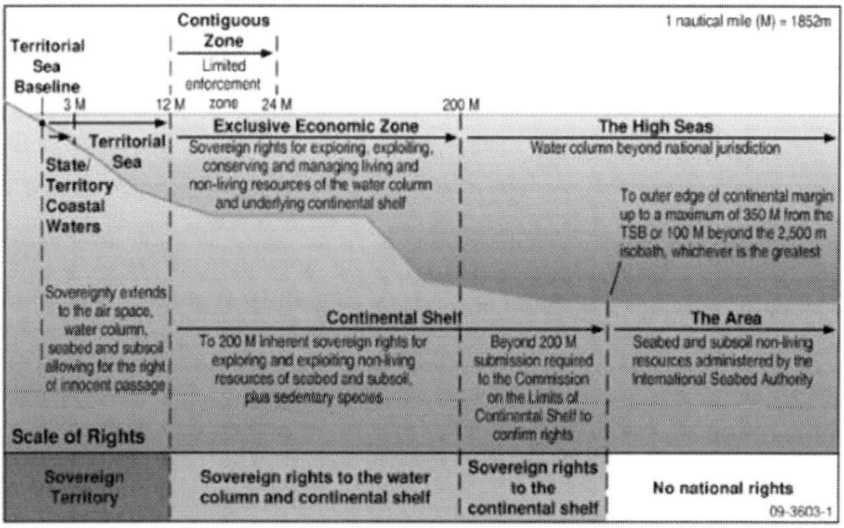

Figure 2.1. Maritime zones and rights under the 1982 UNCLOS.

That is to say, the continental shelf is the first zone of water coming right after the coastal plain ends (See Figure 2.1). The coastal state possesses exclusive sovereign rights concerning the exploitation of its continental shelf's natural resources. Other states cannot explore nor exploits those natural resources without the express consent of the coastal state. Therefore, the coastal state is entitled to lay and maintain submarine cables and pipelines (a system of pipes, often underground, using pumps and valves for flow control, used to transport crude oil, natural gas, water etc. on great distances) on the continental shelf.

Chapter 3

ANALYZING ENERGY SECURITY

…energy supply is a matter of national security.
(George W. Bush 2006)

3.1. INTRODUCTION

This chapter examines energy and security by focusing on the concept of energy security, the factors that affect energy security, the "4As" system, the policy of differentiation, the independent regulatory authorities, the volatile factors and energy security, energy security infrastructure and maritime dimension of energy security.

Energy security could be described as a combination of the concepts of economic, national and environmental security. Economic security refers to the creation of jobs and suitable conditions so that the market for this sector is efficient and secure. National security concerns the safe and orderly functioning of society and the state, especially in terms of energy demand and supply. Environmental security relates with environmental (mainly air) pollution and the degree to which it affects public health as well as with greenhouse gas emissions and global warming.

In addition, economic security is linked to the provision of adequate resources and is therefore a key pillar of the smooth running of the state and the well-being of its citizens. The concept of security, including the terms of energy and economic security has a different interpretation depending on the history and culture of each state. An important role in the process of defining these concepts is played by the identification of national values and interests, elements which are often the result of a common understanding (Boudière et al., 2013).

Environmental security is influenced and defined by factors related to politics, economics, diplomacy, cultural elements, military correlations, the environment and general processes and local and international processes that play a decisive role in the level of protection of the individual, the community the region and , the state during the promotion of their interests. Environmental security is also characterized by profound changes in the functioning of societies. On the one hand, old "players" or agents of the international scene may disappear or shrink, while on the other hand, factors such as states, emerging economies, multinational corporations and non-governmental organizations emerge and play a catalytic role in the global developments. Thus, in the context of globalization, environmental security has evolved into something very fluid and complex, and therefore changes taking place in one part of the world can cause changes in any other place.

Tackling the energy issue is a matter of paramount importance. Rising energy prices and growing dependence on energy imports endangers security and competitiveness. Basic strategies need to be devised to reduce emissions and tackle climate change. In addition, investments must be made to properly shape and cope with the growing needs and new conditions of existing facilities and infrastructure (Ross, 2012).

In the coming years, the dependence on oil and gas will increase. Emerging forces, such as China and India, need increased energy. Mineral fuels will become more expensive and harder to find. Factors that could contribute to the major energy issue are the intensification of discussions on climate change and the future of nuclear energy, as well as threats to energy suppliers from terroristic actions. Energy supply systems are interconnected via automated transmission networks and are vulnerable to cyberattacks. In addition, the political instability that prevails in several exporting countries is a source of concern (Ross, 2012).

For most experts in the field, energy security is directly linked to domestic energy production by the state concerned and minimization of dependence on imports. However, the concept of energy security includes three axes: first, the provision of alternative sources of supply, second, the identification of alternative energy routes, and third, the safeguarding of existing sources and transmission networks. NATO's Strategic Defense Framework describes the protection of vital supply networks as a critical issue for the organization's security and sustainability (Lindley-French, 2006).

Clearly, the described energy security is threatened not only by terrorism, political instability, armed conflict and piracy but also by climatic conditions and phenomena such as hurricanes, floods, earthquakes or anthropogenic disasters. Leading forces and multinationals are resorting to controversial methods to control energy sources, including even military intervention. In addition, various paramilitary groups are acting against governments in order to control the wealth-producing resources and profits they generate. In the future, we expect that once oil production reaches its maximum level of consumption and prices continue to rise, conflicts over these finite sources will intensify, becoming a constant issue for the international community very high on the energy agenda.

3.2. The Concept of Energy Security

Energy Security has long focused on issues such as the reliability of energy supply, the high concentration of energy sources in specific areas, and whether or not, this is a factor in destabilization and the implementation of political energy management tools. Energy is an integral part of any form of economic activity.

The increase in energy consumption has been directly linked to industrialization and economic growth for over a century. With the escalation of energy supply disruptions, discussions on energy security have expanded to include the general reliability of supply chain of electricity, gas oil, as well as the overall energy supply chain, including relapses. A number of different events and phenomena have disrupted supplies, such as political instability in the Middle East and Eastern Europe, natural disasters (the earthquake and tsunami that hit the Fukushima plant in 2011), gas disputes with Russia-which destabilized Europe's energy markets, and blackouts similar to the one that occurred in the United Statesin 2003. Also, Hurricanes Katrina and Rita in 2005 caused, among other things, disasters that affected the energy supply network coming from all sources (oil, gas, electricity) (Diaconu & Rusu, 2013).

Recently, other aspects of energy security have begun to emerge. With the steady rise in energy costs, pricing and its impact on economy competitiveness have begun to be largely considered and play a role in the security of energy supply. Energy price volatility and growing uncertainty about availability of oil and gas imports have highlighted the importance of policies in improving energy efficiency. With climate change and air pollution high on the international agenda, the viability of energy systems has become part of energy security. In case that the environmental conditions worsen to the extent that they affect the smooth running of societies and threaten states, this will have a wider impact on international security.

While energy supply remains a key issue, other issues such as global warming, economic growth, air pollution, economically affordable energy prices and international energy relations are also expected to become a priority, as they will largely determine the degree of a smooth transition to a secure future without energy conflicts and concerns. However, another subject that should be taken under consideration in the long run is whether a broader international coordination of states in the above cases would be preferable to the adoption of individual security policies (Takao & Setoguchi, 2012).

3.3. Factors Affecting Energy Security

The number of factors and the variables affecting energy security, vary significantly. The complexity of the issues that are integrated in the context of energy security is great and varies depending on the perspective on which it has been examined and evaluated. This is one of the reasons why there is no specific definition that corresponds to the concept of "energy security." To make clear how complex the issue of energy security is and the number of factors associated with it, it should be noted that over the last decade, the definitions of energy security within the academic literature abound exceeding the number of 45 (Sovacool, 2011a).

Elkind (2010) states that energy security includes four elements directly related to it: availability, reliability, affordability and sustainability. Availability refers to the ability of users and consumers to have access to energy at any time. Ensuring this requires the existence of an extensive network of buyers, sellers, proper use of natural resources, investment, technology development and a legal framework that will determine the rules of operation of these factors. Reliability relates to ensuring energy supply and its protection from network interruptions. Economic affordability includes stable energy prices that the vast majority of citizens will be able to afford. Sustainability refers to the minimization of social, environmental and economic costs of operating in every link of the energy chain. In addition, the following criteria play an important role: differentiation of the energy mix (in terms of suppliers and forms of energy), differentiation in the supply chain, durability and ability to manage emergencies, reduction of energy demand, proper and orderly operation of markets in this area (Tsaltas, 2006).

Table 3.1. Data and factors mentioned above that affect energy security as well as the complications they can cause (Tsaltas, 2006)

Elements	Components	Threats
Availability	Physical endowment of producers	Exhaustion of reserves that can be extracted cost effectively
	Ability of producers, transit countries, and consumers to agree on terms of trade	Limits on development opportunities such as resource-nationalist policies and state to state contracts
	Technological solutions for production, transportation, conversion, storage, and distribution	Problem in siting infrastructure including NIMBY syndrome
	Capital investment	Financial, legal, regulatory, or policy environments that inhibit investment
	Viable legal and regulatory structures	
	Compliance with enviromental and other regulatory structures	
Reliability	Robust, diversified energy value chain	Failure of energy systems due to severe weather and natural disasters
	Adequate reserve capacity	Failure due to poor maintenance or underinvestment
	Protections from terrorist attacks and political disruptions	Attack or threat of attack by military forces and terrorist organizations
	Adequate information about global energy markets	Political interventions such as embargoes and sanctions

Elements	Components	Threats
Affordability	Minimal price volatility	Exhaustion of reserves that can be extracted cost effectively
	Equitable prices	Energy prices that require lower income households to expend large shares of their income
	Transparent pricing	Excessive subsidies that distort prices
	Realistic expectations about future prices	Failure to institute sound pricing policies
	Prices that reflect full costs	Failure to incorporate enviromental and social costs to energy production and use
Sustainability	Low emissions of greenhouse gases	Adoption and promotion of carbon intensive energy infrastructure
	Minimal contribution to local, regional, and global forms of enviromental pollution	Impacts of indoor and outdoor air pollution associated with energy use
	Protection of energy systems from climate change	Impacts of a changing climate such as ises in sea level, storm surges, and severe weather events

3.3.1. The "4As" System

In recent years, many of publications in the field of energy security refer to the so-called "4As" system (availability, affordability, acceptability, accessibility), which initially appeared as a concept at the Asia Pacific Energy Research Center (APERC, 2007). In essence, the "4As" are factors that directly affect and greatly contribute to energy security. Jewel et al 2014 noted the great similarity of the "4As" to the "5As" in the field of access to health care (availability, accessibility, accommodation, affordability, acceptability) introduced in 1981 (Penchansky and Thomas, 1981).

Energy security's "4As" are frequently mentioned and are a useful tool for modern research in this area. Two of the four elements, availability and affordability, occupying a prominent place in energy security studies (Deese, 1979- Yergin, 1988) and being cited by the IEA in its definition of energy security, are described as follows: continuous availability of energy sources at an affordable price "(IEA, 2014). The other two terms (accessibility, acceptability) are cited as energy factors by the World Energy Council in the Millennium Declaration (WEC, 2000) but they are associated to the concept of energy security only after 2007, when, as mentioned above, the APERC did so.

The four elements that make up the "4As" system are multifactorial. They have many dimensions and are related to many factors of energy security.

Availability is related to the physical and geological existence of energy resources and the ability of a country or region to secure them. The degree of availability depends on a number of parameters such as: (Tsaltas and Deniozos, 2007)

- the ability of supply to satisfy demand (supply security),
- the degree of dependence on imported energy and the extent to which a country or region can rely on its own resources once imported energy is reduced or stopped (self-sufficiency),

- the multiplicity of energy resources and of energy suppliers (diversification of energy sources), the percentage of energy (from the energy mix that is used to satisfy the demand) that comes from RES (Renewable Energy Sources) and
- the overall reliability and potential of a country's infrastructure (technological development).

Economic affordability refers to economic factors having to do mainly with market stability and market liquidity. It depends on parameters such as: (Tsaltas and Deniozos, 2007).

- the impact of energy policies on the market prices and thus on the market stability (price stability),
- the degree of per capita imported energy (dependency),
- the ability of a country to sell energy resources without causing significant price and value fluctuations (market liquidity),
- the extent of usage and geographical distribution of energy facilities (decentralization) and
- the percentage of the population that has access to a reliable electricity network (electricity).

Acceptability depends on the social and environmental costs associated with energy production and use. The degree of (social) acceptability relates to a number of factors such as:

- the environmental consequences (eg. greenhouse gas emissions, water pollution) of the production and use of energy (environment),
- the extent to which energy systems are well-perceived and accepted by the general population (social satisfaction),
- the ability of the national institutions and bodies to regulate the energy sector (national government),

- the degree to which a national state complies with international governance rules, such as the rule of non-corruption (international governance), the use and consumption of energy in an efficient way (efficiency),
- the emphasis placed upon research and development (R&D) aiming at the advancement and evolution of technology relating to energy systems (innovation) and
- the degree of investments made on the energy sector and the relevant job creation linked to these investments (investments and jobs) [Tsaltas and Deniozos, 2007].

Accessibility has to do with the geopolitical dimension of the issue at hand and more specifically with a country's or region's access to external energy sources. Accessibility depends on parameters such as:

- the degree of reliability and credibility of the country's network of energy imports from third countries (fixed imports),
- the quality of relations and policies affecting the international energy trade (trade),
- the level of stability and resilience of a country's domestic institutions and bodies (political stability),
- the security of a state and its ability to safeguard its interests (military power) and the extent of resilience of energy systems to hazards and external threats such as terrorism and natural disasters (safety and reliability) (Tsaltas and Deniozos, 2007).

All the above elements and factors are directly related to and affect the energy security of each state. What is important to keep in mind though, is that, in any policy mix employed with the aim of maximizing the state's energy security, all of these elements play an important role, irrespective of whether some are more carefully considered that others.

3.4. Differentiation Policy

The term energy security is used to refer to the amount of energy a state needs to function properly, that is, to adequately feed its social, economic and military activities. Along the same lines, the term "European Energy Security» can be used to refer to the set of political and legal actions taken by the EU to ensure the uninterrupted supply of the Union with the adequate energy resources for the coverage of its basic energy needs. One of the main objectives of the specific European policy is to prevent the creation of monopolies and / or oligopolies or to reduce the influence of those already existing, on the Union's energy imports (as, for example, is the case with Russia's dominant position on European natural gas imports).

Particularly in terms of natural gas, the EU pursues the above goal with two basic policies. The policy of diversifying its suppliers and the policy of diversifying natural gas transit to its market. Finally, yet importantly, the reference to the existing gas transit capabilities on the territory of the importing countries is considered. Thus, today natural gas is transported from its place of production to the buyers, either through pipelines or through properly designed ships. In the first case, which is the most common today, natural gas is routed through labyrinthine (usually) high pressure pipelines (land or submarine) from its place of production to the importer's territory. This method offers, on the one hand, the ability to import / export quite large quantities of natural gas and, on the other hand, a stable and uninterrupted access for importers. On the other hand, the disadvantages could be the commercial anchors that may be created for a state that cannot differentiate the route of transit of natural gas transported via pipelines to new and possibly more advantageous buyers. This can only be done with the construction of a new pipeline, with all the financial costs involved. As far as the transit of natural gas through ships is concerned, it concerns exclusively liquefied natural gas (LNG). In this case, the natural gas is transported through pipelines from their place of production to terminals, where it is liquefied, after being cooled to -162 degrees Celsius. From there, it is loaded onto special tanker ships and transported to the importers' territory, such as oil. The advantage of this method is clearly the greater freedom of differentiation when tried by the producers of their buyers, without having to deal with the cost of construction of new units or pipelines. As a negative element we could mention the limitations set by the liquefaction process, in terms of the total annual exported quantities of natural gas (compared to the transport possibilities offered by the pipelines), but also the fluctuations in the final price of the product, as well as the risks associated with human (terrorist attacks, piracy, accidents) or natural factors such as weather conditions.

3.5. Theory and Concepts of Geopolitics and International Relations Theory

This section acknowledges the fact that energy security is understood not just as an economic but also as a geopolitical and IR-issue. In that respect, the theories of geopolitics and IR will be explained in order to answer the second sub-question, which deals with the Union's energy dependency towards specific energy partners. Thus, the theoretical tools of geopolitics and IR- theory will be used to analyze more generally the international actors' interests and intentions and the Union's position as a global actor in the field of energy security.

3.5.1. Energy Security and Geopolitics

Geopolitics is more broadly defined as 'the analysis of geographic influences on power relationships' (Bradshaw, 2009). The concept of geopolitics in energy security is referring to the influence of geographical factors, such as the distribution of centers of supply and demand across regions (Bradshaw, 2009). Geopolitical power in the energy sector can be perceived as 'the capability of the state to acquire control of national energy resources and transportation infrastructure and to use or adjust them in the pursuit of foreign and security policy goals' (Siddi, 2018). New emerging energy demanding markets, such as China and India and new developing energy production centers in the Middle East or Africa, affect the geopolitical balance, which results in a global shift. The global shift is driven by different rates of population growth, urbanization and economic development and increases competition over energy supply partnerships (Bradshaw, 2009). In that respect, the EU's position to attain secure energy supply emerged through internal and external developments of geopolitical and economic origin (Crellijé & van der Linde, 2006). In that respect, the internal development concerns 'the process of dual integration', referring to the enlargement process and deepening of the Single Market. Instead, the external development considers the policymaking being influenced by IR and power politics. Along with this distinction, there exist two storylines according to which the geopolitical actors can be characterized. First, global actors can be characterized according to the markets and institutions storyline. It assumes the actor to strive for the continuation and intensification of 9 the international system based on multilateral relations and the globalization of markets. In terms of the future oil and gas industry, the market and institutional storyline reveals the influence of liberalization and market forces issued by the World Trade Organization (WTO), the International Energy Agency (IEA), the Organization for Petroleum exporting Nations (OPEC), as well as regional free trade organizations, such as the EU, the North Atlantic Free Trade Agreement (AFTA). It appears to be the most profitable storyline for the EU. The second storyline is about 'Regions and Empires' and stresses a more pessimistic view on international politics (Crellijé & van der Linde, 2006). Accordingly, the international system is split up into competing blocks that engage in rivalry and competition about energy resources and energy markets. Furthermore, the approach pre-assumes a division into countries and regions based on ideology. Consequently, the storyline stresses a realist understanding of world politics (Crellijé & van der Linde).

3.5.2. Energy Security and International Relations

As already indicated above, energy security plays an essential role in the relations between states. According to Enno Harks (2006), competition over the remaining resources led to the politicization of energy as a commodity and the power struggle in the international system. Power is defined as 'the capability to attract or persuade other international actors' (Siddi, 2018). While the realist school understands power presumably as military power, liberalists associate power with the relative strength of the internal market and regulatory instruments. A third approach of the constructivist school understands power as a socially constructed concept and emphasizes also soft power as means of imposition. In the case of the EU, it can be asked whether the Union pursues a market-orientated and liberal policy or increasingly shifts its focus towards 'Realpolitik' in the context of external energy policy (Siddi, 2018). In that respect, IR-theory cannot be excluded when explaining the relation between states. Realism, liberalism and constructivism can be identified as the major components of this classical theory (Jackson & Sørensen, 2016). These perspectives are further distinguished based on their view on human beings, states and the sources of conflict. Realists acknowledge the relevance of the balance of power between states that are perceived as the main actors. This approach emphasizes the power struggle of states, which aim to survive during a state of anarchy (Jackson & Sørensen, 2016). Liberalism provides a more positive view on human nature and claims that international relations are not determined by a state of anarchy and conflict, but ruled through interdependencies (Jackson & Sørensen, 2016). The constructivist approach emphasizes the 'human's awareness or consciousness and its place in world affairs'. The system that surrounds people is socially constructed and consists of thoughts and ideas (Jackson & Sørensen, 2016).

3.6. Independent Regulatory Authorities

The internal community market requires regulators to ensure open access to the network, protect the right of customers to choose their supplier, and enforce anti-monopoly legislation. Competitive markets require regulators to perform their duties impartially, without discrimination between market players. They must be independent of shareholders and with activities carried out as transparently as possible. In most European countries, various institutions share the responsibility of overseeing the industry. There are three broad institutional approaches to gas supply in Europe as well. First, independent regulatory bodies separated from the ministries have been set up in Denmark, France, Italy, Ireland, Portugal, Sweden and the United Kingdom. These organisms often cover both the gas sector and the electricity sector. In most cases they are responsible, among other things, for setting up the network. They are mainly run by a collective board of directors and operate on the basis of public consultation and other procedures aimed at improving transparency. A second case is for the ministries to handle most regulatory responsibilities immediately. A third group of countries has set up ministerial regulators. They are dependent on the ministry, but are designed to operate autonomously in the management of their day-to-day regulatory affairs. The scope of their activity is similar to that of independent bodies. Directive 2003/55 / EC, all Member States of the European Union were required to establish an independent regulatory authority with a wide range of regulatory obligations, including the regulation of monopolies (unbudling, network pricing and access). terms, rules for system operators), end-user invoices, quality and performance standards, market surveillance and performance monitoring, rules, entry regulation (licensing) and permits), counseling in the respective Government and resolving disputes.

With only a few European countries producing gas, the role of regulators in the upstream gas industry is very limited. However, regulators play a role in facilitating access to markets and entry points both to facilitate investment and to prevent congestion. Another important role is to facilitate the construction of cross-border pipelines / interconnections and liquefied natural gas terminals. At European level, the regulatory authority does not play a role in providing access. The upstream business sector is a matter for Member States with Community guidelines for natural gas downstream.

The role given to regulatory authorities in terms of gas supply security can be very complex, but it does not include defining responsibilities. Their main task is to ensure that effective incentives are implemented. For regulators, the challenge is to ensure the security of gas supply without distorting competition and to ensure the proper functioning of the domestic gas market (Neumann, 2004).

3.7. Unknown Factors and Energy Security

Energy security depends on the ability to obtain an adequate, reliable and affordable supply of energy that provides a degree of independence. Achieving that however is not an easy task, given that there is a long list of geopolitical factors that affect energy security.

Firstly, maintaining the stability of the petroleum and natural gas production countries is of paramount importance for the respective global markets. Ethnic and religious or of other origin conflicts can destabilize a country and disrupt the smooth flow of energy, as in the case of Iraq and Libya. In addition to that, possible asymmetric attacks in energy production and transit infrastructure, by extremist terrorist organizations operating in the Middle East, present a real threat to the stability in the area. In that respect, the proliferation of weapons of mass destruction, especially nuclear weapons, through the black market could have unpredictable consequences if they fell into the "wrong hands" of terrorist organizations.

Apart from that, it is true that the prospects for an embargo on oil-producing countries, such as the one that led to the 1973 oil crisis, seems unlikely today, since the system for managing such crises has improved to the extent that the level of panic is reduced, prices are avoided, and hoarding is maintained. However, the possibility that the international community or a group of states will impose sanctions at an oil-producing state remains real, as evidenced by the sanctions imposed by the Trump administration on Iranian oil exports. Such sanctions exacerbate the overall uncertainty in the global oil market.

Another parameter, potentially detrimental for the energy market and consequently for energy security is the likelihood of a natural disaster. Natural disasters occurred in production, transport and refining areas cannot only be detrimental, but they can also have a big impact on prices and local production, creating the need for imports, as has been the case with the hurricanes that hit the Gulf of Mexico in the past.

It is a fact that the global energy market has the potential to absorb most of the fluctuation arising from local conflicts and internal security issues, as long as there is a significant surplus of energy exports and prices are relatively low. However, market behavior changes dramatically when power supply is limited, and prices are relatively high. In this case, even potential threats to energy production, transit and distribution can radically change prices and market behavior in general.

Yet, real attacks or large-scale industrial accidents could have a much more serious impact. The loss of a major supplier or the prolonged decline in exports due to these factors could potentially have unforeseen consequences, both in terms of prices and supply, thus affecting the entire world economy.

Although in the volatile region of the Middle East, one can see the existence of all the aforementioned volatile factors, the most likely interruptions to occur in the energy flow are short in duration and limited in scope. The shortages created in these cases are mostly addressed by increased production from other countries. The effects of such factors on the economy are usually not greater than those observed due to increases in hydrocarbon prices. The steady rise in global energy demand from the Middle East area, means that the global economy will become increasingly vulnerable to major power outages from the area under review.

But there are also cases that could cause more serious financial side effects. The stability of the major energy producers - Middle Eastern countries - is of great importance for ensuring energy security on the European continent. Any turmoil in countries such as Saudi Arabia or Iraq would cause panic in the international financial markets, as it could affect the smooth flow of energy to the importing countries. Overall, historical fluctuations in the price of "black gold" have shown that a very serious distribution must take place in order to have a significant impact on its price.

However, it is difficult to calculate the psychological impact of a sudden cessation of energy flow in the markets, as it is also clear that the large changes in energy prices caused by the markets have had a real, albeit mild, effect on the real economy. In any case, the economic impact of such an interruption could be mitigated by the timely coverage of demand from countries capable to directly increasing production, such as Saudi Arabia and Russia in the case of oil, combined with the exploitation of strategic reserves (Cordesman & Al-Rodhan, 2006).

3.8. Security of Energy Infrastructure

Energy infrastructure security requires a holistic approach that examines the boundaries, ways and means of locating and exploring energy resources and improving the capabilities of those who store, transport and distribute energy products in the union. In this way, the security of energy infrastructure must cover the entire supply chain from the countries of production and transit in the markets of the importers. In addition, several risks that may jeopardize the proper functioning of the energy infrastructure must be taken into consideration. As key risk factors we would describe natural hazards (i.e., physical safety), security of information and communication technology (ICT), human factors (e.g., negligence, intentional attacks against infrastructure) and some organizational aspects, like the interaction between public and private bodies for emergency response or the negative impact of external assignment to energy infrastructure maintenance. (Borchert & Forster, 2007).

Therefore, energy security is a perfect example of modern security challenges without easy solutions. Energy infrastructure security not only crosses the boundaries of a single nation or region, but also exceeds the responsibilities of national ministries and international organizations, requiring close interaction with non-state actors such as national and international oil and gas companies. In short, the security of energy infrastructure requires a networked government approach. A report on the networking describes the need for close co-operation between public and private entities, at national and international level in all countries in the supply chain. Governance depends on the ability of these actors to implement and maintain their energy infrastructure security policy. This, in turn, requires a common understanding of the challenges, rules and procedures for tackling energy security.

NATO governments (though not NATO as a whole) have already been involved in military-nature efforts to secure energy resources. The first Gulf War, while not a NATO operation, involved key member states such as the United States, France, Britain and Italy, which sought not only to liberate Kuwait but also to ensure that Iraq would not control Kuwait's oil and threaten Saudi Arabia and other Gulf producers. NATO governments also took part in a military operation in the 1980s, explicitly designed to secure oil supplies (Gallis, 2006).

Operation Earnest Will was an attempt, mainly by NATO members, to protect the movement of tankers in the Gulf during the Iran-Iraq War (1980-1988). Beginning in 1984, Iran first, and then Iraq, attacked oil tankers holding a flag of a neutral country, to cut off financial support from the other ruts. Iran attacked Kuwait and Saudi tankers in their own territorial waters, making clear that all Gulf states were not safe.

The Soviet Union, and immediately after the United States, made offers to Kuwait, which lost most of its tankers, to transfer its ships under their flags. An offer that was accepted in the case of US offer. Iraqi planes attacked USS Stark in 1987, killing 37 sailors, since then the US administration has formed a coalition of like-minded states, primarily from NATO members, to protect the flow of tankers in the Gulf. Britain, France and the Netherlands were important participants in Operation Earnest Will.

The Allies seized Iranian boats that were mooring shipping lanes in the Gulf and were exchanging fire with Iranian troops. The operation was not a NATO one, but was an early example of a "willing alliance" focusing to bring security to a strategic area.

In February 2006, NATO governments discussed a range of possible actions in the event of a future cessation of oil supplies caused by military action. Some Member States have repeatedly raised the possibility of protecting oil platforms and the movement of tankers in times of conflict as well as using satellites to monitor developments in areas where energy resources and infrastructure are threatened. Recently, attacks at energy infrastructure sites have multiplied relatively, mainly in Libya, Algeria, Syria and Iraq. Terrorist sabotage of energy infrastructure (EIS) poses a potential political threat to energy security (Weiss et al., 2012).

While a steady, uninterrupted supply of energy is vital to the proper functioning of modern societies, energy transmission infrastructures (ETIs) such as natural gas and oil pipelines, marine tankers or grid pylons and power plants are always almost impossible to be protected.

Given their obvious physical vulnerability and importance, EIS could be expected to be strategically attractive targets for potential attackers. In Europe, governments have implemented Community EIS protection legislation with the issuance of Council Directive 2008/114 / EC. An effective allocation of public resources requires, in addition to recognizing critical infrastructure departments, a better understanding of the type of targets potentially targets by terrorist attacks, ie incentives to attack target categories, including a variety of options.

Analyzing the patterns and characteristics of global terrorism actions against ETIs in the period 1998-2008, Toft, Duero and Bieliauskas showed that these attacks were relatively rare. Geographically, the attacks were extremely uneven, with only three countries, Colombia, Iraq and Pakistan accounting for nearly two-thirds of the world's attacks. There is also no strong correlation between the ideology of the various terrorist groups and the number of attacks on ETIs. Given the basic political ideology of the groups combined with the particular combination of the effects of bullying, the symbolic value and expediency of the attack, and the assessments of outsiders in general, it appears that terrorists have limited incentives to destroy completely ETIs.

If the goal is to disrupt the energy supply, threatening this type of target is not particularly intimidating, it is rarely a strong message of ideological symbolism and, although power transmission networks seem to be soft targets, it requires some skill and knowledge to destroy. In addition, externally interested parties can in many cases be an obstacle to terrorist attacks on energy transfer targets. The incentive framework is not responsible for the unequal geographical distribution of attacks. However, the fact that most of the targets of energy infrastructure attacks were in Colombia, Iraq and Pakistan - countries with high levels of domestic instability and violence - shows that under conditions of violent internal conflict, where groups are conducting military attacks, motives are replaced by tactics of war, where the neutralization of enemy forces through the destruction of power lines (such as energy) becomes a key strategic goal.

In assessing energy security, the low rate of ETIs attacks compared to other types of targets suggests that ETIs are not the main target of terrorist groups. This should be taken into account in the financing of the respective security measures, in order to ensure the energy infrastructure without exaggeration in the security expenses. In addition, when designing new transport infrastructure, it is important to consider whether the supplier or transit country has an increased risk of internal instability and/or internal violence. In such cases, if there are problems and there are no alternative routes, or they are not viable, political/diplomatic efforts to stabilize the situation may be prudent. The driving forces behind terrorism and goal selection are undoubtedly extremely complex (Klare, 2006).

In the ever-changing and unpredictable world of politics, presenting current trends in the future may be unwise. Similarly, depending on the type of technological advancement, the resilience of ETIs could either be improved or become more vulnerable to attack. This may prove to have significant implications for the future of terrorism and the tendency to attack ETIs, as it will change the feasibility of attacks on such targets. In the field of energy infrastructure security, the EU has much to gain from its cooperation with the United States, both in terms of know-how, experience, education, and information, and in the provision of logistical infrastructure that the United States willingly provide to allied states.

However, the example of Iraq and Nigeria must never be forgotten; that any force, no matter how developed it is, will be able to achieve 100% energy security. In the aforementioned cases, terrorist and guerrilla groups have managed to carry out attacks on energy infrastructure, despite the applied military-type security measures and the huge mismatch between the attackers and the security forces (Klare, 2006).

3.9. The Maritime Dimension of Energy Security

The transportation of energy resources plays a crucial role in ensuring regular supply of energy, and thus in energy security .Given the fact that, a large amount of the world's fossil fuel resources are transported by ships, it is of utmost importance to ensure the unhindered movement of ships at sea. However, the number of threats to maritime energy trade, is large and the provision of maritime security to ships at sea is therefore essential.

More specifically, it is a fact that oil provides greater energy security than natural gas, as it is transferred to markets through an internationalized maritime transport system. There is also sufficient infrastructure to store it in the countries of consumption, which allows short-term interruptions in its flow, which is not the case with natural gas. Therefore, the unhindered transportation of oil plays an essential role in ensuring energy security. Given the fact that 60% of the world's oil, a significant volume of which is channeled from the Middle East to Europe, is transported by sea, the development of strategies to ensure maritime energy security has been a priority in recent years (Stampolis, 2019).

This becomes even more necessary with the gradual spread of power, the rise of economic interdependence, changes in geopolitical landscape and the focus of US interest in Asia, which creates new data in the area of maritime security for the EU to deal with potential maritime threats. One of the most important threats is the blocking of the sea routes and the closing of specific sea "choke points", located in the Middle East. which are hugely important for transporting crude oil and liquefied natural gas (LNG) to European markets.

Among the most important sea "choke points" are the Ormuz Straits. Given the fact that about 2/5 of the world's oil is traded through these straits, their closing would require the use of alternative routes, but with higher transport costs. The real ruler of these straits is Iran, which controls seven of the eight major islands in the region and has three naval bases that provide Iran access to the open seas. In the past, Iran has threatened to close the straits if economic sanctions are imposed on it.

Another sea "choke point" of critical importance is the Narrow Bab al-Madeeb (Tears Gate). Oil tanker ships heading west, from the Persian Gulf to the Suez Canal or the SUMED oil pipeline, must pass through these straits. Narrow Bab al-Madeeb is located between Djibouti and Eritrea in Africa, and Yemen in the Arabian Peninsula. It connects the Red Sea with the Gulf of Aden and the Arabian Sea. Closing the straits will prevent tankers from reaching Suez and the SUMED pipeline, diverting them to round Africa, which will increase the cost and time of transit while blocking the availability of tankers. If the El-Madeeb Narrow Babs close, it is possible to bypass the East-West oil pipeline, although the movement of oil from north to south will continue to be hindered.

Unstable Yemen is considered strategically important because it controls the Al Bab Madab Straits on the Arabian Peninsula. Each year, nearly 25,000 ships pass through the straits, amounting to about $ 700 billion in goods and nearly $ 2 billion in barrels of oil.

Finally, yet importantly, one should not undermine the importance of SUMED Canal Pipeline. After the oil destined for Europe passes through the Strait Bab al-Mandeb, it must pass either through the Suez Canal or being transferred into the SUMED pipeline in Egypt in order to reach the Mediterranean. A possible closure of the passage to the Mediterranean will divert the tankers around the southern tip of Africa (Cape of Good Hope), increasing the cost and time of transit, and binding the availability of tankers (Stampolis, 2019).

To face all these threats and to ensure energy security, the EU has developed a European Maritime Security Strategy. This strategy confirms the EU's cooperation with NATO in the field of maritime security, while strengthening coordination between the two (Stampolis, 2019).

Chapter 4

The Eastern Mediterranean in the European Union Foreign and Energy Policy

energy security is our permanent concern as our country's natural resources are inadequate. (Chinese energy official, Xinhua 2011b)

4.1. Introduction

This chapter deals with the European Union's interests, strategy and policy regarding the Eastern Mediterranean region. The European Union deserves a whole section as it is the main actor on the demand side of the Eastern Mediterranean natural gas scene. In other words, the EU holds the potential to balance the region's conflicts and launch, or not, its natural gas market on a continental scale. As a matter of fact, the EU is actively seeking to increase its energy security. According to the International Energy Agency (IEA), energy security is "the uninterrupted availability of energy sources at an affordable price," (IEA, 2018). However, the EU has long been dependent on its Russian natural gas' importations. Being dependent on Russia, the EU risks to undergo natural gas disruptions. It has now become urgent for the EU to diversify its energy supply resources and the recent evolutions on the Eastern Mediterranean natural gas market may allow it to do so. In order to optimize its potential Eastern Mediterranean natural gas' imports, the EU has to find the best routes to channel it to its member states. Among all the probable routes, Egypt stands out as the most interesting option. Indeed, Egypt owns LNG plants that remain idle since the country stopped exporting its natural gas. The use of these plants would allow the EU to enhance its energy security as it would lower its dependence to Russia, decrease the needed infrastructures investments, and it would also grant the EU with more flexibility regarding its natural gas contracts. The idea would be to make Egypt become the crossroad of all the natural gas flowing from Israel and Cyprus, to the European countries. However, the latent atmosphere of conflict in the region questions the feasibility of this project. This chapter will argue that the EU could set more, and better, incentives to develop the Eastern Mediterranean natural gas market and be able to serve its own interests through it. Also, it will show under another angle than on the previous chapter that the region needs to become stable, hence reliable, before the launch of its energy market. The stability of the region appears as the condition to the development of an efficient network of natural gas trade may it be within the region or, on a larger scale, with the EU. Therefore, the first part shows the EU interests in the Eastern Mediterranean region. It will then explain the EU's strategy as well as the tools the EU can use to benefit from the Mediterranean natural gas. Lastly, the second part will develop on the role of Egypt as the platform that could lead the natural gas from its suppliers to the EU member states.

4.2. The European Energy Security Framework

At first, the European Energy and Security Policy Framework will be analyzed, as it constructs the ground for the energy security system's analysis. In that respect, the first section aims to answer the first sub-question: 'What are the objectives, competences and tools with which the

European Union shapes its energy and security policy or strategy?' Therefore, Baldwin's security concept will be used in order to illustrate the way energy security is perceived by the Union.

4.2.1. Competences and Objectives of the European Energy Security Framework

The following part focuses on the analysis of the EU as an international organization 'sui generis.' Generally, the Union cannot be perceived as a state nor as a typical institution, since it owns regulatory power, which directly impacts the national law. However, the Union is not fully sovereign, since the taxing power, main executive competences and external relations tools remain within the member states. This phenomenon can be observed especially regarding the competence distribution between the organizational bodies. In that respect, the relationship between the member states and the Union is determined by supranational as well as intergovernmental terms. According to Baldwin, each state possesses different objectives and demands towards the energy sector, responding to individual energy security problems (Cherp & Jewell, 2014). Therefore, a more general security concept will be applied, corresponding to the impossibility to address all the 28 members' interests and demands towards energy security equally. Thus, following Baldwins conceptualization of 'general security,' there must always be a subject to security. In this case, the EU-member states as 'oil and natural gas-importing industrial nations' function as subject to security (Security for whom?). Although the 28 member states follow different nationalistic interests, given their strategic and natural advantages and disadvantages, the Union shares common values regarding energy security policy. Therefore, special attention will be paid towards the common denominator. Especially concerning the internal development of the energy market, the member states are fully aware of the importance of natural gas and crude oil for Europe's economic growth. Additionally, there is a social factor concerning the secure supply of the energy consumer. More generally, addressing energy security is necessary in order to compete with third states in the global energy market. As observed later on, the Union increasingly strives to strengthen its position as a global actor in the energy security sector (Security for which values?) (Cherp & Jewell, 2014). Accordingly, the Union accentuates the importance of the energy supply security under Junker's Commission, envisioning energy accessibility, affordability and a secure, competitive and sustainable energy market for all Europeans. Since the current policy framework recognizes that energy security is closely connected with the plan of the energy transition, the implementation of an energy security system is directly intertwined with climate policies and the realization of the Energy Union Strategy (European Commission Press Release: The Energy Union: From Vision to Reality, 2019). As Art. 11 Treaty of the Functioning of the EU (TFEU) states, 'environmental protection requirements must be integrated into the definition and implementation of the Union's policies. The Union believes in the potential of renewable energy 'for the reduction of greenhouse gas emissions, the diversification of energy supplies and a reduced dependency on fossil fuel markets (in particular, oil and gas)' (Eurostat: Renewable Energy Statistics, 2019). However, the energy policy and energy security policy objectives are more generally defined in the treaties of the Union. Regarding the field of energy policy, Art. 194 (1) TFEU defines the relevant objectives. In that respect, the objectives determine future legislative and include the improvement of the functioning of the energy market, ensuring energy supply, enhancing the promotion of energy efficiency and developing renewable energy sources and promoting energy cooperation. In that respect, the Commission is an important player and manager of the constitutive legal frameworks. Although the political agenda is determined by the European Council, the Commission is actively engaged in the legislative process by initiating new proposals and determining the draft agenda (Art. 17 (2) Treaty on the European Union (TEU)). However, when it comes to the external dimension of energy security policy, the power to decide on the respective objectives and values of energy security is controlled by the member states. Thus, the competence dispute between the member states and the Union sometimes overshadows real progress regarding the implementation of external and internal energy policies. Legally the Union can be perceived as a 'single person, but it is not based on a single constructive document' (Van Vooren & Wessel, 2014). Moreover, the constitutional basis of the EU is composed of two legal frameworks of equal value, since the adoption of the Lisbon Treaty in 2009 (Art. 1 TEU). While the TEU concerns the general priorities and objectives of the Union, the TFEU issues the competence distribution and principles of law. The competence distribution is managed according to the principle of conferral, which states that competences not explicitly conferred upon the Union, remain with the member states (Art. 4 (1) TEU). However, the principle of conferral is determined by the principals of subsidiarity and proportionality (Art. 5 (1) TEU). Thus, the 'content and form of Union action', which 'shall not exceed what is necessary to achieve the objectives of the Treaties' (principle of proportionality) (Art. 5 (4) TEU). Accordingly, Art. 5 (3) TEU states further that 'areas which do not fall within its exclusive competence, the Union shall act only if and in so far as the objectives of the proposed action cannot be sufficiently achieved by the Member States' (principle of subsidiarity). Regarding the field of energy policy, the Union and the member states share competences (Art. 4 (2) i) TFEU), which means that legally binding acts are issued by the member states and the Union (Art. 2 (2) TFEU). Art. 194 (2) TFEU further states, that the EU-Parliament and the Council are involved as decision-making bodies and should follow the ordinary legislative procedure, except that the treaties provide otherwise. If the treaties provide special provisions, then the Council will decide unanimously after consultation of the Parliament (Art. 194 (3) TFEU). Special provision exists regarding external energy policies (Art. 216, Art. 218 TFEU), involving agreements with third states. The specific provisions can according to 'lex specialis', limit the competences of the Commission to represent the Unions interests externally (Van Vooren & Wessel, 2014). Although the Treaties provide the Union with the general competence to act in the field of foreign security affairs policy (Art. 2 (2) TFEU), the member states should not be prevented from exercising their power. A paragon provides the willingness of the member states to maintain their special energy-mix, which as a result creates greater division of the internal energy market. The third feature of Baldwin's security term asks from which threats the Union needs to protect itself (Security from what threats?) (Cherp & Jewell, 2014). The Union became increasingly aware of the threats to its energy security system in 2014 when the supply disruption dispute between Ukraine and Russia deeply affected the Union. In this breath, the Union identified the issue of supply disruption, and lack of strategic infrastructure as major issue, especially regarding is increasing energy dependence and the lack of an appropriate energy relations network. Thus, the Union further strives to not only maintain their desired standards but also actively engage in the global energy market to push back its competitors. However, the issue of energy security became more complex and deals not only with commercial aspects of energy supply and demand balances. Instead, many aspects need to be considered establishing a more resilient energy security system. In that respect, the following part will depict blow-by-blow the available tools and instruments of the current European energy security framework.

4.2.2. Available Tools of the European Energy Security Framework

As already mentioned above, EU-institutions are obliged to implement the energy policy objectives according to their mandate, which is set out more specifically in the Treaties. In that respect, policy documents and directives can be understood as a tool for change regarding the transition of the current energy security system towards greater resilience. The vision of creating a unified, interconnected, secure and sustainable energy security system has been first set up by the European Energy Security Strategy in 2014 (European Commission Press Release: The Energy Union: From Vision to Reality, 2019). The executive strategy enacted by the Commission and the Council in 28 May 2014 still represents the canon of the agenda for a more resilient energy system. The energy security strategy has been created in the aftermath of the Ukraine crisis in 2014 and responds to the threat of possible energy supply disruptions resulting from political disputes in immediate vicinity (LSE, 2019). Since this policy document still can be understood as an embossing paper, which builds the ground for the development of the current European energy security system, it will be analyzed more closely. Other influential directives and regulation, which followed up on the agenda of European Energy Security Strategy, will be discussed in the realm of the strategic goals, proposed by the eight pillars of the European Energy Security Strategy. This procedure allows the identification of follow up measures the Union has been taken based on this policy document. The strategy presents eight pillars, which can be categorized based on their timely purpose (Communication of the European Commission to the Council and the European Parliament: The European Energy Security Strategy, 2014). Thus, the EU perceives certain measures especially urgent to realize a more resilient energy security system. The short-term measures include first extended capacity building, establishing an extensive infrastructure, as well as increasing energy relations with neighboring countries (short-term measures). The short-term measures are of preventative nature and meant to help the Union to maintain its economic, social and political standards. However, other priorities, such as moderating the energy demand, the development of a better integrated single market, the goal of energy self-reliance, the development of new energy technology, the reduction of dependency through diversification, the energy transition to low carbon economy and speaking externally with a unified voice, are perceive as long-term projects (mid to long-term measures) (Communication of the European Commission to the Council and the European Parliament: The European Energy Security Strategy, 2014). Generally, the measures are supposed to increase the overall flexibility of the system and to reduce the relative supply dependence. Consequently, the eight pillars mirror the priorities and strategic aspects of the Union perceives most important in order to create greater resilience against possible energy threats.

4.2.2.1. Capacity Building

Generally, the Union strives to cope with possible energy shocks in the short run. Therefore, the focus of the energy security strategy lies in the adaptation to the fast-changing environment (Communication of the European Commission to the Council and the European Parliament: The European Energy Security Strategy, 2014). Thus, the possibility of immediate response requires the EU to enhance its storage capacity. Additionally, the EU strives to enhance the development of bi-directional gas flow pipelines, enhancing the interconnection of the member states. Further, the EU-Energy Security Strategy stresses the launch of energy security stress tests (Communication of the European Commission to the Council and the European Parliament: The European Energy Security Strategy, 2014). The strategy also advises the Commission and the member states to cooperate more intensively with the Gas Cooperation Group (GCG), which plays an important role in managing the security of gas supply measures between the member states. The GCG has been installed in the breath of the Security of Gas Supply Regulation (994/2010). The group is composed of national authorities, such as the Agency for the Cooperation of Energy Regulators (ACER), The European Network of Transmission System Operators for Gas (ENTSOG), the Energy Community and representatives of industry and consumer associations and coordinates gas supply disruption measures (European Commission). According to the Energy Security Strategy the GCG-network should be extended, due to its important function as a platform to exchange expertise and coordinate relevant information about supply issues among the member states (Communication of the European Commission to the Council and the European Parliament: The European Energy Security Strategy, 2014). In that respect, the Directive 2009/119/EC has been enacted and contributes to short-term capacity building. Based on the directive, the member states are committed to maintain a minimum of crude oil reserves and petroleum products to withstand sudden shocks. Accordingly, a minimum requirement of 90 days consumption is identified as critical buffer in emergency situations (Communication of the European Commission to the Council and the European Parliament: The European

Energy Security Strategy, 2014). These preventative measures have intensively been issued by the Commission's Communication on short-term resilience of the European gas system in October 2014 in response to the supply disruption issues with Russia and the Ukraine and the annexation of Crimea. In that respect, the ability of the European member states to react to short-term disruptions has also been emphasized by the Preventive Action Plans and Emergency Plans (Regulation 994/2010). The emergency planning is further based on the cooperation between the EU and ENTSOG, which exerts an advisory function.

4.2.2.2. Protecting the Strategic Infrastructure and Establishing a Support System

However, the strategy also emphasizes the prevention and preparation to sudden disruptions, including the protection of strategic infrastructure and the collective support of the most vulnerable member states. First, (Communication of the European Commission to the Council and the European Parliament: The European Energy Security Strategy, 2014). The protection focuses especially on energy transmission systems. The EU notably discusses here the interference of non-EU actors, such as state-lead companies, national banks and key suppliers (Communication of the European Commission to the Council and the European Parliament: The European Energy Security Strategy, 2014). In that respect, ENTSOG has been established in order to facilitate the exchange and cooperation between national gas transmission system operators (TSOs) across Europe. According to the new Security of Gas Supply Regulation, ENTSOG is also deemed to perform EU-wide gas supply and infrastructure disruption simulation in order to alert for serious shortcomings (EUR-Lex: Regulation (Eu) 2017/1938 of the European Parliament and of the Council, 2017). However, creating a strategic infrastructure network is essential in order to guarantee cross border supply, but without a proper solidarity mechanism the internal energy security structure is deemed to fail. Thus, secondly, the goal of creating more solidarity between the member states has been stressed by the European Energy Security Strategy and Art. 194 TFEU. Accordingly, the solidarity mechanism refers to a distribution mechanism of energy, which guarantees a minimum level of intra-EU deliveries of energy supply. The principle is essentially contributing to the stability of the internal energy system (Communication of the European Commission to the Council and the European Parliament: The European Energy Security Strategy, 2014). In that respect, the coordination via the GCG-network is increasingly considered and advised also by the European Energy Security Strategy.

4.2.2.3. Moderating the Energy Demand

Furthermore, the EU strives to moderate the energy demand, which is perceived as the most efficient tool to reduce external energy dependence in the long-term. Considering the Energy Efficiency Directive (EFD) and Energy Performance for Buildings Directive (EPBD) significant energy should be saved. However, the savings depend on the priority sectors as well as the mobilization of capital. According to the European Energy Security Strategy, the industry and building sector possess great energy efficiency potential. This mechanism acknowledges further the inclusion of the private sector, which is still perceived as an important investor for the energy sector. However, low-carbon investments are also supported by the European Structural and Innovation Funds (ESI). Nevertheless, the achievement of the 2020 energy efficiency target, which focusses on heating and insulation in the building and industry sector has been particularly emphasized by the EU- Energy Security Strategy (Communication of the European Commission to the Council and the European Parliament: The European Energy Security Strategy, 2014). Furthermore, the moderating of energy demand is closely connected with the implementation of the European Energy Union strategy, which has been set out in the Commission's Communication on a Framework Strategy on a Resilient Energy Union in February 2015. The Communication on the Energy Union emphasizes the 'energy efficiency first' principle (EUR- Lex: Regulation (Eu) 2017/1938 of the European Parliament and of the Council, 2017). The Communication recognizes that climate and environmental policy can never be entangled from energy security matters. In that respect, the temporary policy agenda is also influenced by the 2030 policy framework for climate change and energy framework, adopted by the European Council on 24 October 2014. The policy strives to get rid of at least 40 percent in greenhouse gas emissions, increase the share of renewable energies in energy consumption, improve energy efficiency and develop greater electricity connection (Gouardères, 2018). Another tool includes the 'Clean Energy Package For All Europeans' (COM(2016)860), which aims to keep the EU competitive when the clean energy transition affects the global energy markets. Generally, this package is encompassed by eight legislative proposals that cover 'governance, electricity market design (the Electricity Directive, Electricity Regulation, and Risk-Preparedness Regulation), energy efficiency, energy performance in buildings, renewable energy and rules for the regulator ACER (Gouardères, 2018).

4.2.2.4. Integrated Energy Market

The fourth pillar encourages the development of a fully integrated internal energy market in the long-run (Communication of the European Commission to the Council and the European Parliament: The European Energy Security Strategy, 2014). In that respect, national policy decisions regarding bilateral energy agreements, the investment in infrastructure projects, such as the Nord Stream I and II projects greatly affect the internal cohesion of the market. Furthermore, there is still a high market segmentation regarding the natural gas sector (Johnson & Derrick, 2012). Therefore, this pillar also issues the risks of unilateral energy security measures jeopardizing the proper function of the internal energy market (EUR-Lex: Regulation (Eu) 2017/1938 of the European Parliament and of the Council, 2017). Thus, strategic decisions need to be coherent and exerted at the EU level, not at the national level. Accordingly, the Union aims to stick to a regional approach for the integration of the European energy market. In that respect, integration projects, such as the North West (Pentalateral-Forum) should be further enhanced. Moreover, the ENTSOG-network is meant to complete the internal market for gas and stimulate cross-border trade (Gouardères, 2018). The Energy Security Strategy nevertheless recognizes a lack of well-integrated markets in the Baltics. A common energy market may not only need a shared regulatory framework but also an energy transport infrastructure. In that respect, the regulation on the guidelines for the trans- European energy network and the Connecting European Facility (CEF) should be extended (Communication of the European Commission to the Council and the European Parliament: The European Energy Security Strategy, 2014). Since the European oil market is increasingly dependent on Russian crude oil for the refinery industry and transport sector, the Energy Security Strategy asks the member states to reduce their oil dependencies to Russia especially in the transport sector (Communication of the European Commission to the Council and the European Parliament: The European Energy Security Strategy, 2014). Thus, alternative measures, such as the reduction of greenhouse gas emission and consumption of transport fuels will be supported. With respect to the fourth pillar, the EU aims to strengthen the regional cooperation, complete the transportation of internal energy market legislation and the development of alternative fuels infrastructure (Communication of the European Commission to the Council and the European Parliament: The European Energy Security Strategy, 2014). The Third Energy Package, the Regulation on Guidelines for Trans- European Energy Infrastructure (Regulation (EU) No 347/2013), the Regulation on Wholesale Energy Market Integrity and Transparency (Regulation (EU) No 1227/2011), the Electricity Directive (COM(2016) 0864), the Electricity Regulation (COM(2016) 0861) and the Risk- Preparedness Regulation (COM(2016) 0862) are some of the main legislative instruments meant to contribute to the better functioning of the internal energy market (Gouardères, 2018).

4.2.2.5. Increasing European Energy Production

The fifth pillar encourages a more active role of the EU when addressing the discussion on the production of energy by the member states themselves (Communication of the European Commission to the Council and the European Parliament: The European Energy Security Strategy, 2014). The pillar focuses on measures which can be achieved in the medium or long-term and include the investment in renewable energy, nuclear energy as well as sustainable energy production. Accordingly, renewable electricity and heating can already reduce natural gas consumption in several sectors. These measures should profit from the national and ESI- Funds under the coordination of the EIB and international financial institutions support (Communication of the European Commission to the Council and the European Parliament: The European Energy Security Strategy, 2014). The exploitation of conventional resources in European production areas, such as the North Sea, Mediterranean Sea and Black Sea will be developed in Commission to the Council and the European Parliament: The European Energy Security Strategy, 2014). The security strategy admits that a more detailed revision of EU unconventional reserves (economically recoverable resources) need to be done. In that respect, the development of the European Energy Union plays a crucial role and cannot be disregarded when enhancing energy efficiency and self-reliance to reduce external energy dependency. However, increasing European energy production is strongly related to the development of new energy technology encouraging the transformation of the Union towards a renewable energy Union.

4.2.2.6. Development of Energy Technology

The sixth pillar emphasizes the development of new energy technologies (Communication of the European Commission to the Council and the European Parliament: The European Energy compliance with the environmental legislation (Communication of the European Security Strategy, 2014). Without the investment in energy technologies, energy dependency cannot be reduced in the medium and the long-term. New technologies are needed to reduce the primary energy demand and can offer efficient and cost-effective solutions (Communication of the European Commission to the Council and the European Parliament: The European Energy Security Strategy, 2014). Therefore, the EU and its member states are deemed to invest in research and innovation allowing for the diversification of energy resources. Therefore, the Commission will use the Horizon 2020 Framework Program for research and innovation (2014-2020) and push further the integrated Roadmap of the Strategic

Technology Plan to be consistent with the Energy Security Policy. The European Strategic Energy Technology Plan (SET-Plan), which has been adopted by the Commission on 22 November in 2007 strives to drive the market introduction and integrate low-carbon and efficient energy technologies (Gouardères, 2018). However, the development of new energy sources generating technology can only be achieved in the long run. Nevertheless, this pillar is especially crucial regarding the shift of demand and supply of the global energy market. In that respect, the Union needs to remain an attractive out-led market for energy products. Encouraging a green energy transition can result in additional attractiveness of the market but might also lead to less competitiveness.

4.2.2.7. Diversification

The seventh pillar stresses the diversification approach regarding external suppliers and energy partnerships while maintaining significant volumes from reliable and secure external sources. The diversification of sources, suppliers and structure is perceived as a long-term project. In that respect, the diversification of gas suppliers through Liquified natural gas (LNG) agreements is encouraged by medium to long term measures. LNG is a form of methane that has been converted to liquid form for ease of storage or transport. 'As a liquid, LNG takes up around 600 times less volume than gas at standard atmospheric pressure' (European Commission: Liquefied Natural Gas). Thus, this transformation enables the transport of natural gas over long distances without the actual use of pipeline infrastructure (European Commission: Liquefied Natural Gas). In that respect, the strategy regards the LNG suppliers from Northern America, Australia, Qatar as a promising source. Besides the goal of strengthening the relationship with external suppliers, the EU will more open to new sources (southern Corridor gas delivery form Azerbaijan, identified projects of common interest). Thus, especially suppliers form the Caspian region are attached (Communication of the European Commission to the Council and the European Parliament: The European Energy Security Strategy, 2014, p. 16).

4.2.2.8. A Unified External Energy Position

The eighth pillar strives to improve the cooperation of national energy policies and to speak with a unified voice when it comes to an external energy policy decision (Communication of the European Commission to the Council and the European Parliament: The European Energy Security Strategy, 2014). In that respect, the Union faces two issues. First, decisions on energy mix are a national prerogative. Secondly, the negotiation of energy related agreements with third states is a matter to the member states and their bilateral relation with energy partners. Due to the shared competences in the field of energy policy and the single competence of the member states regarding external energy security relations.

4.3. THE EU'S ENERGY STRATEGY IN THE EASTERN MEDITERRANEAN REGION

Even though the EU has slightly reduced its needs in natural gas importations (The share of natural gas imports fell from one percentage point between 2016 and 2017), the EU imports of natural gas in a gaseous shape still represented 20% of its total energy imports in 2017. 69% of its natural gas is imported. The EU has a long tradition of being highly dependent on its Russian natural gas imports. In 2017, the EU has imported 37% of its natural gas from Russia. Concerning the Russian natural gas dependence, not all EU member states are equal. In 2017, the most dependent states are Bulgaria, the Czech Republic, Estonia, Latvia, Austria, Poland, Romania, Slovenia, Slovakia and Finland, which all import more than 75 % of total national imports of natural gas from Russia. The dependence of those countries is explained by their proximity with Russia. There are also Russian natural gas dependent countries in the Western Europe. Germany imports between 50% and 75% of its natural gas imports from Russia, Italy and the Netherlands both import between 25% and 50% of their natural gas imports from Russia. Germany, Italy, and the Netherlands are followed notably by Spain and the UK, which both import less than 25% of their natural gas imports from Russia, (Eurostat, 2018). The Russian natural gas dependence is a problem for the EU as it gives too much political leverage to Russia. Russia has the power to disrupt the natural gas flow from the EU, which directly threatens the EU members' energy security. The energy security is, as defined by the International Energy Agency (IEA), "the uninterrupted availability of energy sources at an affordable price," (IEA, 2018). The long-term way to install energy security is to invest in order to reach an energy supply that will both promote economic developments and meet sustainable environmental needs. The short- term way to ensure energy security is the ability to deal with sudden changes in the supply or in the demand. The purpose of the energy security concept is to give some tools to avoid negative socio- economic impacts of an energy disruption or unavailability for instance. That is why, it is strongly recommended to vary the sources of energy supply. Therefore, the EU's energy security strategy consists of:

- Developing the energy efficiency in order to reach the proposed 2030 energy and climate goals.
- Increasing energy production in the EU and diversifying energy sources, energy supplier countries and routes. The EU wants to further deploy the renewables sector, as well as sustainable production of fossil fuels, and a safe nuclear energy where this option is chosen. Moreover, there will be negotiations with the current main energy partners such as Russia, Norway, and Saudi Arabia. The EU will also negotiate new energy contracts to diversify its energy sources with countries in the Caspian Basin region for instance.
- Meliorating the internal energy market and creating the needed infrastructure links to be able to react quickly to supply disruptions and to be able to reconduct energy to where it is needed.
- Informing the EU member states on the new developments and agreements of the EU with non- EU states. Also, the EU wants all its member states to speak with one voice when it comes to external energy policy. This is a tough goal to reach as, as

mentioned earlier, all EU countries do not have same energy needs or dependence level, therefore, interests are different from one country to another.

- Developing the solidarity between EU member states and an EU common strategy if there is an emergency to protect critical infrastructures. There should be more coordination between EU countries to use existing storage facilities, develop reverse flows, conduct risk assessments, and put in place security of supply plans at regional and EU level, (EU, 2018).

Hence, putting a hand on the Eastern Mediterranean natural gas market is completely in line with the EU energy security strategy. The European indigenous gas production (from Russia and Norway) is declining but the demand is rising. Concretely, Europe is eager to add new sources of natural gas supply through the South Gas Corridor ("Fourth corridor") in the Caspian Sea for the Azerbaijani and Turkmen gas, as well as through the "Fifth corridor" in the Eastern Mediterranean. The EU aims at first importing 10bcm of natural gas per year by 2019-2020, then up to 100bcm of natural gas in the future, (Tilliros, 2018). The EU will need the Eastern Mediterranean natural gas to be added to the Caspian gas in order to diversify its resources sufficiently.

Also, the LNG is forecast to become competitive enough to take a much larger part in the European natural gas resources. The use of LNG would highly increase the EU energy security. Moreover, it would allow the EU members to benefit from the global LNG market. Indeed, it is important for the EU's energy security to avoid being dependent on pipelines only. Pipelines creates high dependency on the long- term between the exporters, the transition countries, and the importers. In such a relation, trust is necessary. Pipelines therefore multiply the risk of gas disruption because of political dispute, terrorism, or commercial disputes with any of the numerous states located on the pipeline's way. LNG would give the EU access to better competitivity and flexibility regarding natural gas trade. Additional natural gas resources are essential for the EU's economic development on the long-term. The Eastern European countries are suffering from the Russian natural gas dependence an such an event would allow them to further develop economically and to catch up the Western European countries' economy.

When it was Cyprus' turn to obtain a place among the European Union member states, the EU had to deal with the peculiar features of the split island. As a matter of fact, the EU does not recognize the Turkish Republic of Northern Cyprus, which is only recognized by Turkey. However, the EU recognizes the Republic of Cyprus, located in the Southern part of Cyprus, as the only legitimate government of the island. This is problematic as the Republic of Cyprus only controls the Southern part of the island, while the government of the Turkish Cypriots, the TRNC controls the Northern part. The Republic of Cyprus is an EU member state since 2004. The EU regards the northern part of Cyprus as de jure, an EU member state as it considers it to be a part of the Republic of Cyprus. The accession of the Republic of Cyprus to the EU was seen as an opportunity for the island to be reunified. When a referendum on the Annan plan in 2004 asked both communities whether the North and the South should be brought back together, seeing it as an opportunity to reach the EU, the Turkish community voted in favor of it. However, the Greek community voted it down.

The Annan plan was a strategy to reconcile the Northern part of Cyprus with its Southern part. This plan was brought by the, at the time, UN Secretary General Kofi Annan. The strategy developed by this plan was to create the United Republic of Cyprus covering the whole island with the exception of the British Sovereign Base Area. That is to say that Cyprus would be ruled by a federal government composed of two constituent states: The Turkish Cypriot state and the Greek Cypriot state. The federal system was imagined on the basis of the Swiss federal model. There would have been a collective presidential council in which members would be allocated according to the percentage of the state's population. A president and a vice president would have been appointed by the presidential council. This system also included a bicameral structure: a senate and a chamber of deputies, and there would have been a supreme court composed by an equal number of Turkish Cypriot and Greek Cypriot judges. The refusal of this plan by the Greek Cypriots was a huge disappointment for the European Union, (Smith, 2004). The Republic of Cyprus' president Tassos Papadopoulos emphasized that the Greek Cypriots did not vote against the Turkish Cypriots' community, but rather against the Annan plan as a solution to the separation of the island. As the EU had agreed on the Republic of Cyprus (RoC) becoming an EU member regardless of the votes' results, the RoC st indeed became an EU member with nine other countries on the 1 of May 2004. On the protocol N°10 on Cyprus, the act concerning the accession's conditions of the 58 Republic of Cyprus and the nine other countries to the EU, the European Union affirms its wish for the RoC's membership to the EU will benefit all Cypriots citizen 20 and will favor civil peace and reconciliation . This document states that the Cyprus has entered the EU as a whole, only the acquis is suspended for the Northern part of the island, which is not ruled in effect by the government of the Republic of Cyprus, ("Protocol n°10 on Cyprus," 2003). Translated in practice for the daily life: Turkish Cypriots possess, or are eligible for EU documents and are EU citizens. However, the TRNC is not under the EU law, only the RoC is. This situation is frustrating for the Turkish Cypriot community who has voted for the Annan plan but is now left at the doorstep of the European Union. Also, the Republic of Cyprus has hardly met the requirement for its accession to the EU, mostly thanks to the Russian investments on the island. This decision installed confusion among some EU elites, who showed regrets to have let a split Cyprus enter the EU, (Kyris, 2015).

Nonetheless, the EU's commercial relations in Cyprus are quite strong, Intra EU trade accounts for 45% of Cyprus' exports, most of them heading to Greece (12%) and the UK (8%). When it comes to imports, 74% come from EU member states, namely Greece (21%), and Germany (17%). Italy comes third with 7%, (2018). Politically, the Republic of Cyprus is represented by 6 members of the European parliament. Also, the RoC held the Council of the EU presidency from July to December 2012. In 2016, the EU spent €0.185 billion in Cyprus, which represents 1.06% of the Cypriot gross national income (GNI). The same year, Cyprus contributed to the EU budget by €0.152 billion, which represent a contribution to the EU budget of 0.87% of its GNI, (2018).

The EU says that it is willing to set incentives for both parts to take steps towards each other, but for now, it seems like the RoC would have more to lose than the TRNC in the case of a reunification. In order to be efficient in solving the Cypriot conflict, the EU should empower the TRNC and rebalance the power in the region. Even in terms of finances, the cost of reunification would be tremendous, especially for the Turkish Cypriot community, since it doesn't yet benefit from the same standards of living as the Greek Cypriot community living in the RoC. For example, the average per capita income of those living in the north is $12,7019 , while the Greek Cypriots in the Southern part of the island have an average income per capita over twice that at $25,233. 571 . The TRNC still depends heavily on Turkey for financial, political, and military support. Another potential solution to appease tensions on the island and to unlock the situation regarding the maritime rights of both parts, would be to give up on a potential reconciliation and start working with the RoC and the TRNC two separate states. Recognizing the TRNC would be a way to better control its actions. Furthermore, it would install its legitimacy as a state and allow the TRNC to escape a bit of Turkey's influence.

The EU's EEZ is composed by the sum of its member states' EEZ. Each member state continues to exercise its power on its own EEZ. Nonetheless, there exist shared competences such as the control of fisheries on European EEZ, which is exclusively controlled by the EU. Therefore, the boat of a certain EU member state, can legally, under the EU law, sail into another member state's EEZ. The whole surface of the current EU's EEZs represents approximately 20 millions of square kilometers, ("Israel, Cyprus, and Greece push East Med gas pipeline to Europe,"). This figure makes the EU's EEZs the worlds' largest EEZ. Also, the EU's maritime zone is 380% larger than the whole EU land territory (See Figure 4.1).

The maritime space is important for the EU as the maritime trade represents 40% of the total EU trade. Therefore, it is in the interest of the EU and of its member states to cooperate together. It would be interesting for the EU to define a stronger and more precise common maritime law and maritime strategy.

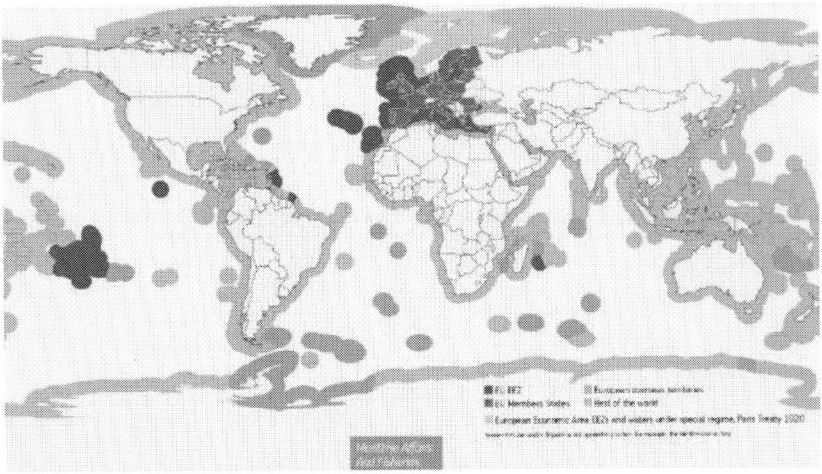

Figure 4.1. The European Union's total Exclusive Economic Zone ("Israel, Cyprus, and Greece push East Med gas pipeline to Europe," 2018).

Concretely, the EU has a way to positively affect the Eastern Mediterranean region notably through the Union for the Mediterranean (UfM). The UfM is composed of the 28 EU member states and of the 18 Southern and Eastern Mediterranean countries. The UfM was created to reinforce the cooperation between the organization's member states and the regional dialogue. Also, it aims at being a platform to incentivize and push the creation of regional projects to meliorate the citizen's lives. These goals are encompassed by three bigger objectives: human development, stability, and regional integration, (UfM, 2018). Concretely, the UfM is composed of three platforms: The UfM Regional Electricity Market platform, the UfM Renewable Energy and Energy Efficiency platform, and the UfM Gas platform. The UfM gas platform's purpose is to structure the necessary dialogue in the Eastern Mediterranean region to better develop its natural gas market by promoting security, transparency, and predictability of changes in demand and supply. All of it, in order to maximize the interests of both the producing and consuming countries.

There are already the same kind of organization in the Western Mediterranean namely, EU- Algeria and EU-Libya. Therefore, the UfM can further focus on the Eastern Mediterranean natural gas sector. In fact, the UfM Gas platform could contribute to the development of a regional cooperation in this area, (Energy: a shaping factor for regional stability in the Eastern Mediterranean?, 2017).

The EU also disposes of the EU Neighborhood Policy, launched in 2004 to support stability, security and prosperity in Southern and Eastern parts of the EU. In order to appease tensions, the EU set the objectives of supporting the regional economies' development and scaling up security cooperation with countries in the area. Therefore, energy security is at the core of the issue. More precisely, the EU aims at offering cooperation on the basis of case-by-case assessment, to give a solid th framework for production, distribution, trade and energy efficiency. On May, the 4 2016 in Washington was held a US- EU Energy Council. On this occasion, the council declared that it "recognizes the potential of the new gas resources in the Eastern Mediterranean for the energy security of the EU and the wider region." It also "stressed the need to respect the sovereignty and sovereign rights of EU Member states to explore and exploit their natural resources and stands ready to facilitate the development of these resources and corresponding infrastructure, underlining the need to respect international law" .

Nonetheless, apart through a few rare and careful statements notably when the Turkish ship had intercepted the Italian vessel off Cyprus , the European Union does not seem willing yet to take a stand in the Cyprus conflict. As a matter of fact, in the beginning of 2018, Turkey sent a vessel off the Cypriot coasts to prevent an Italian vessel belonging to the firm ENI to drill in this zone, the EU warned Turkey to not threaten Cyprus. In fact, Turkey told ENI it had booked the area for military activities. According to the Republic of Cyprus, this area was part of its EEZ. As explained in the previous chapter, Turkey does not recognize the existence of EEZ, and estimates that it is its rights to conduct any activities in this area. Turkey immediately called the Greek Cypriot government out for its "unilateral drilling," which, according to Turkey, undermined Turkish

Cypriots' right to dispose from the island's natural resources. Meanwhile, the new Greek Cypriot president Nicos Anastasiades stated, one week after his election, that Turkey had violated the international law and that the government would take the "necessary steps," (Maurice, 2018). Nonetheless, the president added that he did not want to add fuel to the fire regarding the conflict between the two communities. One of the reasons behind the EU's silence and refusal to take a stand may be that Greek Cypriots were not benevolent in the Annan plan. Also, we can assume that this period of so called "refugee crisis" granted Turkey with some political leverage over the EU countries such as Italy, France, and Germany, all three being founding members of the EU itself.

CONCLUSION

The development of the Eastern Mediterranean natural gas industry and its exportation to the EU countries through Egypt are subject to the condition that there are further natural gas findings in the region. In addition to this, the Cypriot and Israeli conflicts need to be appeased. These are the conditions to attract investments in the region. A successful development of these energy findings would benefit the 65 EU by providing it with intra-EU long-term gas supply, which is helping with the "natural gas resources diversification" part of the EU's energy security. Also, the natural gas potential of the region holds the potential for the EU to open a dialogue with these conflicting neighboring countries. Lastly, this chapter confirms that, as far as the Eastern Mediterranean region is concerned, regional stability on the basis of political and economic discussions is more likely to generate the necessary energy development than the development of the energy market is to generate regional stability.

Chapter 5

ENERGY SECURITY IN THE EASTERN MEDITERRANEAN

The politics of a State lies on its geography.
(Napoleon Bonaparte)

5.1. INTRODUCTION

There is a growing global interest for the Eastern Mediterranean region, which plays a more important role in world politics today than it did at the beginning of the post-Cold War period. The discovery of hydrocarbon deposits in Eastern Mediterranean created new dynamics in the field of geo-economics and can act as a catalyst for stability and cooperation or for confrontation and conflict. On the other hand, in the aftermath of the interruption in Israeli-Turkish relations in 2010 and the Arab Spring uprisings, the geopolitical scenery in the Eastern Mediterranean has changed and new strategic partnerships have been formed. The most important change is that Greece and Cyprus, two European Union (EU) member states, have developed special and separate trilateral partnerships with Egypt, Israel, Palestine, Lebanon, and Jordan. These cooperation schemes include various levels of political collaboration, from the level of Presidents and Prime Ministers to the level of Ministry Directorates. This chapter examines the situation in the basin of the Eastern Mediterranean. In particular, it expands on the European energy security, the Russia-Iran-Turkey subsystem, the Azerbaijan-Turkey geopolitical axis, the Israel-Cyprus-Greece and Egypt-Cyprus-Greece geopolitical axes, the US superpower system, the Russian superpower and finally it analyzes the role of Greece in strengthening national and European energy security.

5.2. European Energy Security

Forecasts for global energy consumption in the medium-term show that the energy mix will continue to consist mainly of oil, coal and natural gas. In fact, the consumption of all three fossil fuels will continue to grow in the coming years. RES will contribute less than 10%, with an annual increase of 1.5%. The paradox is that the proven global oil and natural deposits are constantly increasing (Koppa, 2017).

As a result, it is still possible to produce large amounts of hydrocarbons in the medium term. However, what matters is not when the hydrocarbon reserves will run out. Small deposits or deposits that are not in the interests of exploitation may be discovered for several more years. The most important issue is when production growth will no longer be able to meet growing demand. During that time, there will be a possibly rapid, economic crisis, and even if there are still huge fields available for exploitation, the International Energy Agency (IEA) defines energy security as “the uninterrupted availability of energy resources at an affordable price.” The following are indications of European energy security.

In 2015 Europe produced 5.6% of the world energy and consumed 11.9% of it, while the Middle East produced 13.7% of the world energy and consumed only 5.1% of it. The great energy demands of the “old” continent are met by imports, many of which from the Middle East. The largest share of the world energy production is in oil and its derivatives, solid fuels and natural gas, the total percentage of which was 81.7% in 2015. There has also been an increase in the production of all forms of energy - except for the nuclear one, which is gradually declining - which reflects the growing global demand for energy.

RES, despite the boost in their production, did not exceed 13.2% of the world production. Europe relies heavily on energy imports to meet its energy requirements. These imports increased by 22% from 1995 to 2015, despite a slight decrease in imports between 2010 and 2015. 99.2% of these imports had to do with fossil fuels. The largest increase was recorded in the import of natural gas, which raised to 89% from 1995 to 2015. As a result, the EU's total energy dependence on imports showed an upward trend, from 43.1% in 1995 to 54% in 2015. This is a paradox, if we consider that the overall energy consumption in recent years showed a slight downward trend (Koppa, 2017).

The reasons for the decline in energy consumption should be sought in the deindustrialization that has taken place in recent years in Europe, the improvement of energy efficiency and the increase in energy prices. The realization of the fact that the uncontrolled consumption mainly of fossil fuels significantly degrades the environment and contributes to the greenhouse effect, thus undermining the quality of life of future generations, has pushed the EU to take measures to limit emissions of CO2 and other gaseous pollutants (GHGs Emission: Green House Gasses Emission). In this context, the EU's policy on transition to clean energy forms is being endorsed, promoting the development of RES. In this way, the EU covered 17.5% of its energy needs using RES in 2017 and set a goal to cover 20% of them by 2020.

In parallel with the use of RES, the EU is attempting to disengage from "finite" imported energy from fossil fuels. The wholesale price of electricity in European countries, although low, is still 30% higher than in the United States. Tax imposition significantly increases the final energy price at consumer level. The wholesale prices of gas exceed twice that of the United States. The increased price of energy compared to other economies has serious consequences for the competitiveness of industrial sector. The EU is investing heavily in the energy sector. EU-based RES companies are among the most developed in the world, as they are profited by EU policy on disengagement from imports and the development of competitive technologies to produce clean energy. (Energy Union Package, 2015).

5.3. Iran-Turkey Subsystem

Russia has developed close relations with Iran and Turkey in several strategic sectors. In the commercial/economic sector, exports from Turkey to Russia increased by 59.1%, in 2016 compared to 2015, reaching $ 150 million, while imports from Russia to Turkey increased by 10.2%, reaching $ 1.5 billion. According to data released by the Turkish Statistical Institute, the largest share in imports from Russia is related to energy, reaching $ 832 million, followed by steel-iron products, which amount to $ 279 million.

In the energy sector, Turkey undertook the nuclear project in Akyu, which amounts to 22 billion dollars. That has prompted the German secret services to make presumptions about the existence of a secret Turkish program intended towards the development of an atomic bomb. Of course, such a task is not easy to carry out because it depends not only on technical capabilities, but also on the existence of a huge financial budget, as well as on the willingness of the Great Powers and Israel to allow such a thing. However, Russia is not only involved in Turkey's nuclear program, but also in Iran's nuclear program, which is under the control of the "Five Plus One" Agreement, which is now being challenged by the United States, colluding with Israel. In addition to the nuclear program in Akyu, Turkey and Russia are cooperating on several other energy projects, such as the Turkish Stream (Balodimos, 2016).

Last but not least, in the military sector, one of the most important issues under consideration relates to the sale of the S-400 anti-missile system from Russia to Turkey. This system is a "sophisticated weapon", which affects the balance of power between Turkey on the one hand and Greece and Cyprus on the other, in favor of Ankara. Russia has installed S-

300s in Iran as well as S-300s and S-400s in Syria. This evolution affects the military balance in the region, especially between Israel, Egypt and Turkey as well as the British military bases. This is mainly becausethese missiles and the S--400 radar have a long range capable to locate and trap targets to a distance up to 600 km. Behind these strategic movements, one can discern Moscow's attempt to weaken Egypt's position and strengthen Turkey's position in the area. The reason is simple. The Egyptian regime derives its international legitimacy from the United States and any moves by Turkey will affect the foreign policy and interests of the US. There is also the following view: The tighter the cordon around Israel, the greater the importance of Cyprus for Tel Aviv as the only way out of the logic of strategic depth (Balodimos, 2016).

In that direction, Russia aims at the weakening of Israel's position that works for the benefit of Turkey. Israel accepts and supports the Egyptian regime because considers it more friendly and credible than that of the Muslim Brotherhood. Therefore, by weakening Israel's position, Russia helps Turkey to meet its goal, which is to consolidate its leading role in the Muslim world.

In the same direction, any attempt of Turkey to create problems in the relations between Egypt and Cyprus is supported by Russia. Under the current circumstances, Egypt and Cyprus have a bilateral agreement to delimit their overlapping (exclusive economic zone) EEZ entitlements in the Mediterranean. Both countries are interested in maintaining this agreement, especially now that there are increased energy interests in the area. Morover, Egypt knows that it can trust Cyprus much more than Turkey and it prefers stable relations with Israel rather than a relationship with Ankara, which will be of dubious intentions. Given that, Ankara's permanent objective is the annulment of the Cyprus-Egypt agreement for the boundaries of the EEZ and all its attempts to create problems in the Egypt-Cyprus relationship point in that direction. (Balodimos, 2016).

5.4. Geopolitical Axis Azerbaijan - Turkey

Strong ethnic ties between Azeris and Turks have marked the relations between the two countries throughout history. The people of Azerbaijan are Turkish-speaking and the official language of the country is a variant dialect of the Turkish language, enriched with Iranian elements. The close relations between the two countries have continued to grow in recent years, even after the declaration of independence of the "Republic of Azerbaijan".

The factors influencing the proximity of the two countries are the perpetuation of the Nagorno-Karabakh conflict, Russia's close contacts with its declared adversary, Armenia, as well as the agreements on the transfer of Azerbaijan's Caspian energy wealth through the territories of Turkey. In addition to their common origin, the two countries share common interests in the Caucasus region, maintaining privileged relations as well as to engage in foreign policy (Boudière et al., 2013).

5.5. Geopolitical Axes Israel - Cyprus – Greece and Egypt - Cyprus – Greece

According to the US Geological Survey (USGS) conducted in 2010, the Levantine Basin, one of the 8 geological - basins of the Eastern Mediterranean region, which covers the territorial waters of Cyprus, Israel, Lebanon and Syria, is estimated to have 3455 bcm of natural gas reserves, which are comparable to Iraq's reserves. Recent discoveries of large hydrocarbon deposits have redefined the energy landscape of the Middle East, increasing the proportion of natural gas in the region's energy mix to 27% over the period 1990-2009. Israeli natural gas discoveries Tamar and Leviathan in 2009 and 2010 respectively, however, were the ones that transformed the Levantine basin into an area that could export gas in the future (Pavlović et al., 2018).

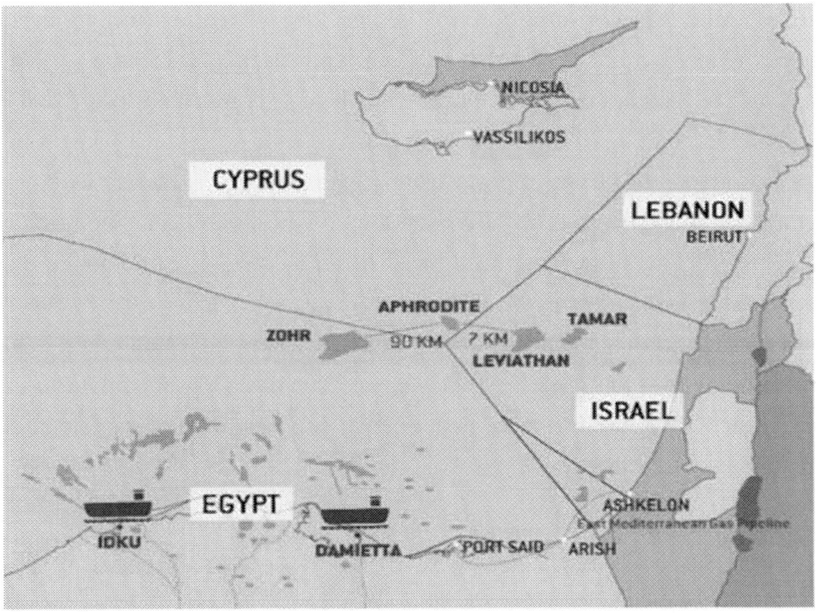

Figure 5.1. Deposits in the Eastern Mediterranean (Cyprus - Egypt - Israel) (Bornstein, 2018).

The total stocks already discovered account to 515 bcm, a number that is considered insignificant in relation to the world sizes, but recent discoveries in the Cypriot and Israel territorial waters are estimated to reach 1076 bcm with expectations of growth in the near future.

Since 2009, the regional energy balance in the Eastern Mediterranean has changed radically. In addition, its geographical location at the border of Middle East energy sources makes it of equal importance to the Caspian and East African regions (Bornstein, 2018).

5.5.1. Israeli Deposits

Israel is one of the largest consumers of natural gas in the Eastern Mediterranean. Gas consumption has increased in recent years, with an annual average of 0.0099 bcm in 2002 reaching a peak of 5 bcm in 2011 which is expected to rise to 9.7 bcm in 2015 and reach 13.3 bcm in 2020.

Much of the gas consumption in the last decade came from domestic sources (Mari-B field discovered in 2000 and reached 4.25 bcm in 2012) but Israel relied also on gas imports from Egypt to cover up to 40% of its domestic demand in 2010 (Chorev, 2018).

High dependence on imports from Egypt poses a high risk of supply disruptions as repeated attacks on sections of the Arab Gas Pipeline have taken place. The submarine section of this El-Arish - Ashkelon pipeline carries natural gas from Egypt to Israel.

The situation changed drastically in 2013 when the Tamar field began producing natural gas yielding a proven reserve of 283 bcm of natural gas. The larger will be used to meet Israel's ever-growing domestic gas demand (Dagoumas & Flouros, 2017).

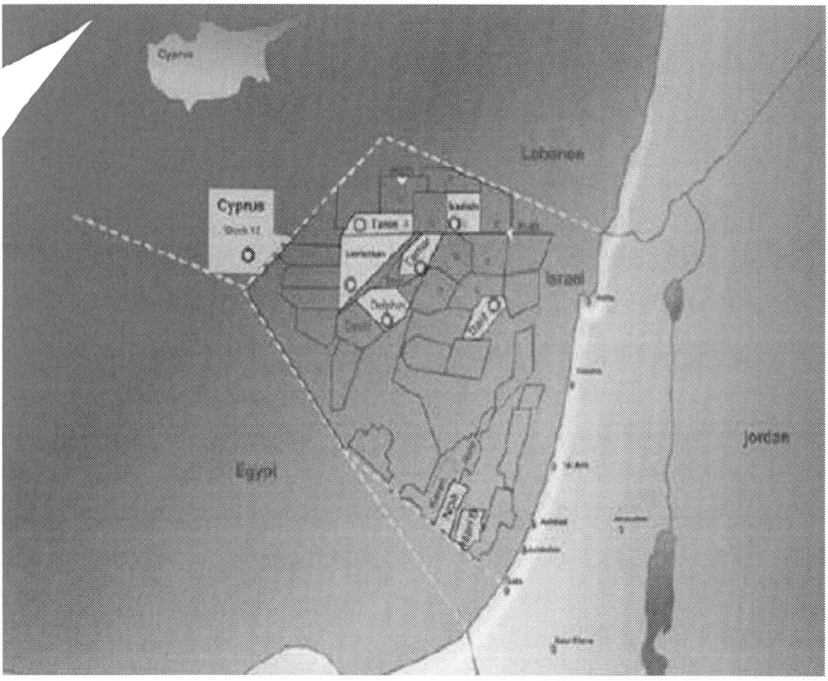

Figure 5.2. Deposits in the Israeli EEZ.(Chorev, 2018).

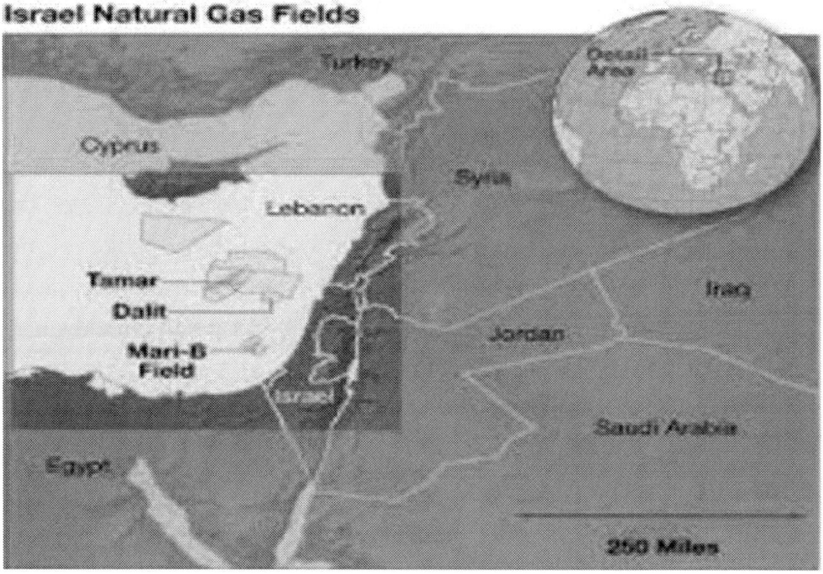

Figure 5.3. Deposits in the Israeli EEZ (Elkind, 2017).

Estimates based on work on the Leviathan gas field performed by the consortium Israeli Delek (45%), Ratio Oil Exploration (15%) and US Noble Energy (40%), indicate that this field could generate 510 bcm, but production is not expected to begin before 2017. All the natural gas discoveries have dramatically improved Israel's energy profile. The Israeli Ministry of Infrastructure, Energy and Water Management encourages the transition to natural gas consumption as the main energy product of the domestic energy mix in view of its economic and environmental benefits. The share of natural gas in electricity generation increased from 20% in 2008 to 40% in 2010 with the support of the state-owned electricity company, while overall gas consumption is expected to increase by 50% by 2015 (Elkind, 2017).

In June 2013, the Israeli government confirmed its decision to export 40% of its gas production (about 410 bcm), with first possible destinations being Egypt and Jordan, as supplies are expected to outpace demand despite criticism from the opposition. At the same time, in view of the prospects for industrial development based on the increase in gas supply but, recognizing the need to enhance energy security, Israel will seek to counterbalance excessive dependence on this gas, by preventing natural gas infrastructure and LNG facilities) more vulnerable to terrorist attacks.

Asked how Israel sees things in the Eastern Mediterranean, Israel's Professor Uzzie Aranda said at this time of turmoil: "the political forces are stronger than the economic and religious strongest of all. Israel aims for future conquests and defending its resources." He spoke of new protagonists in the region, Russia and Iran who are on the agenda of winning a seat in the Eastern Mediterranean. "Our goals should not be grandiose but temperate so as not to lead to losses and disappointments. The area is complex and requires prudence, careful and gradual approach."

5.5.2. The Cypriot Deposits

The coverage of Cyprus' domestic needs is almost exclusively based on oil and electricity, while gas consumption is at a zero level. After Malta, Cyprus has the least diversified energy mix of the 28 EU member states.

In November 2008, US Company Noble Energy was licensed to explore hydrocarbons in Venus Block 12 after its 10th licensing round in 2007. The Aphrodite gas field is 34 kilometers west of the Israeli Leviathan Field, most of it is within the Cyprus EEZ. Cyprus will have 80% in case of co-operation with Israel. The exploratory drilling took place between September and December 2011. At the end of 2011, Noble Energy announced the discovery of a reserve capable to deliver 142 - 226 bcm. In February 2012, the Republic of Cyprus announced its 2nd Authorization Round by publishing the relevant announcement in the EU Official Journal (İşeri, 2019).

As an EU Member State, Cyprus is subject to the EU legal framework applicable to the hydrocarbons sector. Hydrocarbons (Search, Research and Production) Law 4 (I) 2007 incorporates European Directive 94/22/EC on the conditions of use of licenses for the prospecting, exploration and production of hydrocarbons in Cypriot national law. In addition, regarding Directive 2009/73/EC on internal market rules for natural gas, Cyprus has secured an exemption from the commitments regarding third party access to the natural gas network as Cyprus is an isolated market gas that until recently did not even have natural gas. This exemption will cease when Cyprus is not an emerging gas market announced the 2nd Licensing Round by publishing the relevant announcement in the EU Official Journal (Kontos, 2019).

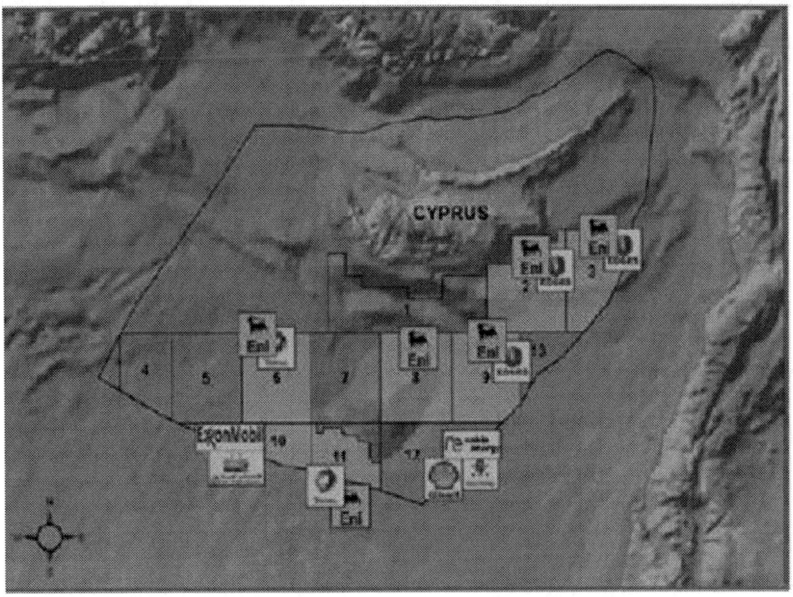

Figure 5.4. Fields to be exploited in the Cyprus EEZ (Kontos, 2019).

Exploration of hydrocarbons within the Cyprus EEZ is of vital economic importance, as it can make a significant contribution to the country's energy independence given the current economic crisis. Exploration agreements between businesses and the government of Cyprus should be based on Production Sharing Contracts (PSCs) for which a bonus payment must be paid upon signature. These bonuses could amount to more than $ 200m in direct revenue for the Cypriot state.

Panos Kelamis, Managing Director of Cyprus Hydrocarbons, said that "everything starts and ends with tanks - without stocks we have nothing." He stressed that gas prices are impossible to predict and pointed out that the planet needs a lot of energy that is interconnected with prosperity. "The energy mix of the future should focus on fossil fuels. Hydrocarbon exploration is the new El Dorado for gas." Finally, he said that the Republic of Cyprus is carrying out a strategic plan and that it will have 2 oil wells in 1.5 years (Philokyprou et al., 2017).

5.5.3. The Egyptian Deposits

Egypt is the second largest producer of natural gas in the Southern Mediterranean, and across the African continent after Algeria. The Arab Spring had a negative impact on Egyptian gas exports to Israel, Syria, Lebanon and Jordan mainly because of the terrorist attacks against the Arab Gas Pipeline. The political turmoil of 2011 and 2012 brought production to a standstill, which was at 62 bcm a year, and as a result it now faces difficulties in meeting the ever-increasing domestic gas needs. Gas accounts for 49% of Egypt's total energy mix with upward trends (Mondal et al., 2019).

Current estimates indicate that Egypt has 2180 bcm of exploitable natural gas reserves. About 70% of the overall Egyptian production comes from the conical offshore deposits of the Nile Delta. The natural gas exported, reaches third countries either through pipelines (30%) or in the form of liquefied natural gas (LNG) [70%]. Cairo has three major liquefaction plants, SEGAS LNG Train 1 and Egypt LNG Train 1 and 2. These plants can produce an amount of LNG that could reach 16.8 bcm per year.

The increase in domestic gas demand is expected to result to a decline in LNG exports. In 2012, 53% of total Egyptian LNG exported was to Japan, followed by North Korea, India, Taiwan and China. 39% of Egyptian LNG exports were absorbed by the European market and Turkey. In concrete terms, France, Portugal, Spain and Italy introduced Egyptian LNG. Before 2012, European markets were the main destination for Egyptian LNG exports. In 2012, there was a 25% decrease in export levels to the EU compared to previous years. This decline reflects the general trend of the lowering of LNG imports into the EU and of the increase of LNG competitiveness on world energy markets (Sartori, Lorenzo & Paceviciute, 2016).

The prospects for producing additional quantities of natural gas from the Nile Delta area are however very positive and BP and German RWE have recently announced the discovery of exploitable deposits. The prospect of co-exploiting the deposits located in fields 4, 5, 10 and 11 with Cyprus is also important as these deposits are in the EEZs of both countries. The issues of maritime delimitation, however, are the subject of an analysis of the following sections (Asvestopoulos, 2018).

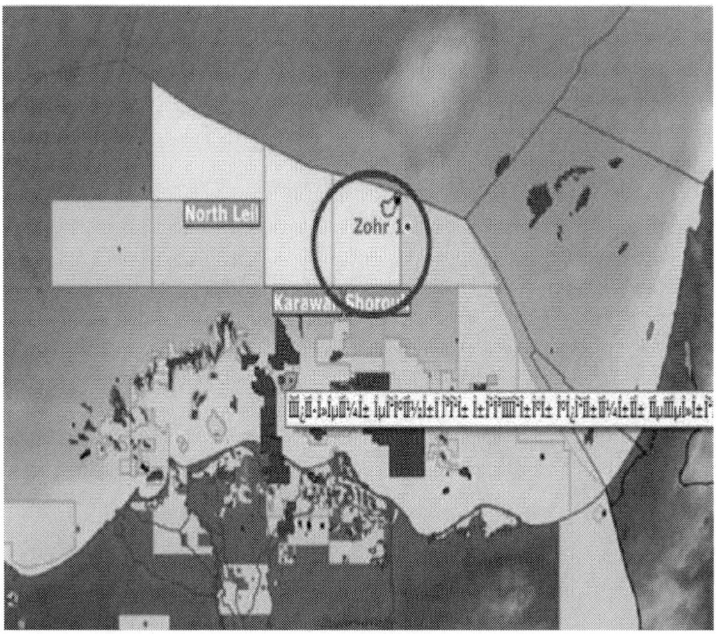

Figure 5.5. Deposits in Egyptian EEZ (Sartori et al., 2016).

5.6. The US Supersystem

For the US, the Mediterranean Sea is "the place where the Gulf begins" as well as an important waterway for the access to areas of possible conflict and for the transfer of energy resources. As long as the situation continues to be as it is today, their interest in the Mediterranean Sea will be undiminished. It is a fact that whoever dominates the Mediterranean, lays the foundations to dominate the whole world. For many years during the last three decades, the Eastern Mediterranean has been an effective tool for US national security. The strategic position of the Eastern Mediterranean in the middle of Europe, Asia and Africa, has made it the epicenter of great competition for power since antiquity. As a result, it acted as a breeding ground for the United States to take its first steps as a global superpower at the beginning of the Cold War. Eastern Mediterranean seems to have the largest discovered deposits of coastal natural gas with prospects for significant additional findings. This potential benefit for the region is overshadowed by the fact that many regional agencies have competitive territorial claims on these valuable resources, leading to the increase of both the existing tensions and of the the risk of conflict. However, in the midst of these crises, there are also opportunities for the United States such as improving our allies' cooperation, which could help consolidate regional stability and weaken Russia's energy efforts in Europe. In addition, the United States is interested in other issues in the region, such as the fight against terrorism, drugs, human trafficking, and immigration. As it has been recently shown, Greece is also willing to invest heavily in NATO and its bilateral relations with the United States. However, while US policy makers have taken significant initial steps to support and promote such cooperation, there is much disagreement over how to handle relations with Turkey. This reflects a wider uncertainty about US intentions and commitment. Turkey, under President Recep Tayyip Erdogan, is no longer a reliable member of NATO as it's becoming increasingly authoritarian and unpredictable, unlike its former partners - especially the United States and Europe, but also Israel and Egypt. Russia's acquisition of advanced S-400 air defense is indicative of the fact that Turkey's traditional role as NATO's southern stronghold is being abandoned (Tsakiris, 2018).

5.7. The Russian Supersystem

Russia's interests in the Mediterranean have a long history. Prior to the time of Peter the Great (late 17th and early 18th centuries), Russia had no ports in the Black Sea and, as a result, little interest in Mediterranean issues. Until the end of the 18th century, Russia undertook campaigns to create passages in the Aegean, the Black Sea and generally throughout the Mediterranean. These actions resulted into conflict with Britain. As of the beginning of the 21st century, a new Russia has emerged. A Russia that, although not exactly an ally, is no longer the enemy, threatening the West. It has recently become a state that shows signs of confidence and strength. Making full use of these characteristics, Russia is cleverly maneuvering in the geopolitical arena, with surprising moves and decisions intended to establish itself as a global player.

For Russia, the Mediterranean offers ports to "warm waters" and an outlet into the high seas. Another one of Russian's long-standing dreams has been the control of the Straits. The Mediterranean Sea is an area in which Russia is trying to revive its naval presence, which, as the past has shown, has contributed to Moscow's diplomatic development. In general, Russia's policy evolution in the Mediterranean Sea has a fourfold targeting. First, Russia sees North Africa and the Middle East as a "land bridge" for access to the entire Muslim world. Secondly, Moscow is pursuing a separate agenda while dealing with southern EU members, rather skeptically maintaining the European Union's ability to speak with a single voice in some cases where developments in the Mediterranean are seen as a continuation of the Kremlin's challenges in the Black Sea. Moscow's foreign policy is driven by its desire to regain in political power over the United States and by its insistence that it would be treated as a major power as a counterbalance to the global influence exerted by the US. At a first glance, the Mediterranean seems to provide fertile ground for such a strategy. Meanwhile, there is a sense of over-simplification in Russia's reasoning: in the tradition of zero-sum mentality, the Kremlin tends to say that "dethronement" of the United States would automatically lead to the strengthening of Russia's position.

In addition to its military aspirations, Russia is pursuing policies aimed at an economic gain. Thus, the Russian economy can benefit from geo-economics in two ways (Tsakiris, 2018). The first is through its participation in the exploitation of the existing deposits on the Syrian coast. This site is in the Eastern Mediterranean basin and in addition to its geostrategic role, it can generate geo-economic profit capable of stimulating Russia's economy, which is largely based on the energy resource trade. The second is through the promotion of the Russian defense industry not only to the Syrian regime but also to other Mediterranean states such as Turkey (which is also involved in the Syrian war with the recent attack called "Source of Peace"), and Libya. This is how the Russian defense industry is stimulated, ranking it among the three superpowers in this field.

5.8. The Strengthening of Supranational Actors

In recent years, the relationship between national and supranational factors seems to be changing. The supranational networking and law has now begun to strengthen its power over the national one. There is a tendency to weaken state power, a tendency to regulate international markets at the supranational level and an expansion of multinational corporations. On the other hand, of course, there are the corresponding resistances that are just as strong.

The strengthening of national perception, the revival of protectionism in cases where the safeguarding of national interests, monetary competition and the policy of regionalization of geo-economic zones is required, are typical examples of the reaction to the tendency to strengthen supranational institutions. The complexity of this ongoing competition is variable and redefined. In conclusion, it should be noted that, in the current era of postmodernism, neither one state can be the absolute sovereign, nor can it possess complete national independence, let alone economic self-sufficient. The relations of interconnection and interdependence of states are now diverse and numerous.

In this sense, in the current world regime, the weight of the traditional nation-state in the decision-making process tends to be lower than that of other, mainly international actors, such as organizations, coalitions and consortiums of economic interests. The new reality includes the overcoming of national autonomy, the global spread of the free market, the global movement of capital, investment, technology and goods (Tsakiris, 2018).

The new Strategies of the petroleum companies and their objectives to restore dominance in the production sector based on the above framework. The strategy of the multinational companies operating in the field of energy, has as its ultimate goal the return to the old regime where they fully controlled the oil industry and they distributed only a small percentage of the profits to the governments of the producing countries, consequently absorbing significant power from these states.

In fact, US and UK oil companies have always sought to control the global market through secret agreements with the producing countries. The result of these agreements was the maintenance of high selling prices and, of course, the high profitability of producing companies in the Middle East, through the exploitation of low-cost deposits in the region. Due to the very high profitability of the assignment of low-cost oil fields, these types of assignments are not made on the basis of the current market. On the contrary, the most profitable deals are made through the political and military power of the governments that control the deposits.

In modern times, the major energy-producing countries of the Middle East have created their own, national, energy exploitation companies. Their main goal is to replace multinational companies in energy production and maximize their profits. The multinationals, for their part, claim that they have all the tools needed to make more efficient and effective use of fossil fuels than the troubled nationalized companies that often face internal turmoil, organizational problems, corruption and lack of capital and technology to research new deposits (Mazis, 2008).

5.9. Collisions and the Geopolitical "Black Hole"

5.9.1 Libya

Libya's influence on Europe's energy dynamics is extremely important, and it highlights a significant triptych of energy interests set against the implementation of EastMed's interests in Germany, Turkey and Russia. Germany and Russia have already reached an agreement on the well-known Nordstream consortium of pipelines passing through the Baltic Sea, starting with the already running Nordstream pipeline, which supplies gas to the EU. They are still investing in Nordstream II, which has not yet been completed. It will be completed while it has to deal with the reactions of the United States, which considers it a trap for the EU. Germany, through Nordstream II, wants to make it possible to receive quantities that will pass through Poland and Ukraine. Germany looks forward to 2020 to the success of this project (Tsakiris, 2018).

Through Libya, Turkey aims at the weakening of Greek's position so that it cannot play a decisive role in the region, not only in the field of energy but in other fields as well. Finally, Russia is taking advantage of Libya to use it to control the supply of hydrocarbons from the southern Mediterranean.

5.9.2. Syria

The non-military intervention in Syria by the Obama administration, despite the French government's willingness to provide military assistance, has shaken Washington's prestige and credibility among its Arab allies, especially Saudi Arabia. Saudi Arabia was mainly the one that, along with Israel and Turkey, had attempted to overthrow the Assad regime in order to dismantle the Shiite axis that supplied Hezbollah from Iran. Also, with the fall of the Damascus regime, not only would Iran be cut off from the Mediterranean, but its influence in Iraq would be limited and Hezbollah's threat to Israel would gradually be reduced. However, the withdrawal from the forefront of the Americans, the French and the British created a geopolitical power vacuum that allowed the Russian military to intervene. It is worth noting that Syria is a prime example of the modern form of proxy war, in which Iran, Turkey, the United States, Saudi Arabia and Qatar clash over military and paramilitary groups they support each one, without directly involving themselves (Tsakiris, 2018).

5.9.3. Iraq

However, the early withdrawal of US troops from Iraq was followed by the unpredictable appearance of the Islamic State in the country's Sunni triangle. Obama also stepped down from Iraq, underestimating the influence that Tehran had built on Baghdad's now-dominant Shiite political spectrum. In fact, the presence of American oil companies is now limited in the country, with the main representative ExxonMobil, which is active in increasing production in the West Qurna 181 field (Tsakiris, 2018).

5.9.4. Yemen

Iran's intervention through the Shiite Houthis in Yemen has provoked a backlash from Riyadh and its coalition, as Yemen shares land borders with Saudi Arabia and controls the strategically important Bab al-Mande Strait. Saudi Arabia also fears that expanding control of Houthis in Yemen could encourage the country's Shiite minority to challenge the Saudi royal throne. It is noteworthy that Saudi Arabia and Israel have put aside their differences and joined forces to prevent Iran from gaining more power and political influence in the wider geographical region.

The United States, Britain and France, which are permanent members of the United Nations (UN), are providing diplomatic support to the government of Aden, which opposes this Sana'a, the Houthis. An additional concern for the United States is its involvement in the Yemeni, al-Qaeda and Islamic State militant groups. The evolution of the coalition's operations has not yielded the expected results (Tsakiris, 2018).

Aden's forces, centered on Saudi Arabia, have come under fire from the international community for the civilian casualties they have caused, mainly through the airstrikes they carry out. This criticism extends to the unprecedented food crisis in the country, as a result of prolonged hostilities and the coalition's efforts to cut off all supplies from the Hutis.

5.10. US Economic Sanctions on Iran

On November 5, 2018, US Secretary of State Michael Pompeo said: "More than 20 countries have already cut off oil imports from Iran, reducing Iranian exports by one million barrels a day." This is just one of the negative effects of the Trump administration's renewed economic sanctions on the Iranian economy. As a result of the above development, there has been a sharp decline in Iranian oil exports between 2017 and 2018, which has raised concerns about the energy security of countries dependent on Iranian oil. The following eight countries were excluded from the restriction on Iranian oil exports: China, India, Japan, South Korea, Taiwan, Turkey, Greece and Italy.

Saudi Arabia and Russia have rushed to fill the gap created by Iran's oil exports, as they have the potential to boost production. However, there has been an upward trend in the price of black gold, as has always been the case in similar cases. In a statement, US Treasury Secretary Steven Mnuchin said: "The imposition of unprecedented economic pressure on Iran by the Ministry of Finance should make it clear to the Iranian regime that it will face increasing economic isolation and economic stagnation until its behavior changes fundamentally ..."

The move by the United States is aimed at hurting Iran's missile and nuclear programs, as well as reducing the Islamic Republic's intervention in the Middle East. They are believed to be the reason for Washington's withdrawal from the agreement on Iran's nuclear program, on May 8,

2018, an agreement in which the EU continues to commit itself. The effects on the country's economy are already serious and will continue to decline during 2019-20. Inflation in December 2018 was beating at a rate of 34.9% and there is an increase in prices for basic necessities. The re-imposition of sanctions has reduced confidence in the country's economy, resulting in a currency crisis, while the local currency has lost much of its value.

The country's government is trying to develop an Economy of Resistance. Signs of popular unrest are evident and the government is preparing to use violence to control possible outbursts of popular discontent. Nevertheless, Tehran is using harsh rhetoric against the West and even the United States, and is showing no signs of changing its foreign policy as tensions continue to escalate. At the same time, it is using the regional crisis that is plaguing it to mobilize its citizens.

5.11. Italy Is Reviewing Its Position on TAP

The Southern Gas Corridor is an ambitious project that aims to help detoxify Russian gas imports. That is why it has become the subject of political rhetoric, while Russia, which is threatened by this project, is reacting with the design and construction of new competitive pipelines such as TurkStream. In recent months, TAP has threatened not to become operational, not because of Russian interference, but because of the revision of the Italian government's position. In particular, the Italian parliamentary elections of March 2018 brought to power the Eurosceptic party of Giuseppe Conte. Shortly after the election, the new government's environment minister, Sergio Costa, said TAP's work was futile and was being revised, as Italy aimed to reduce gas consumption for environmental reasons (Andrei, 2015).

The above statements about TAP, by the Minister of Environment, were made shortly before the statement of the Italian Prime Minister at the Brussels summit, at the end of June 2018, where he expressed Italy's opposition to the automatic renewal of sanctions against Russia. Such a "geopolitical" consultation in Europe could link the Italian government's position to its possible decision to stop supporting TAP, which is being built with EU support and is a competitive project in Russia's interests in Europe. Behind Italy's move may be purely economic incentives aimed at cheaper Russian gas, and the prospect of expanding the Russian TurkStream pipeline to the Italian coast would have the same environmental impact as TAP (Andrei, 2015).

Italy currently imports about 90% of the natural gas it consumes, with Russia being the main supplier (51%), followed by Libya, Algeria, the Netherlands and Norway. After all, Italy is not the only EU country that relies heavily on Russian gas. A typical example is Germany, which imports 70% of its natural gas from Russia through the North Stream I and II pipelines. However, a geopolitical approach seems to be expressed by Rome in rejecting the TAP project, through which the Italian government distances itself from the EU and its partner Azerbaijan, for the benefit of Russia.

The government of Rome could really use TAP as a statement of its desire to become independent from Brussels. But the irony is that the Russians may want to save the project themselves. This is because, for the time being, Azerbaijan has the capacity to supply only half of the gas that will be transported via the Southern Gas Corridor (6 billion cubic meters / year destined for Turkey and 10 billion cubic meters / year are intended for European markets) (Andrei, 2015).

As a result, Gazprom is unlikely to be able to channel the missing gas through TANAP and TAP, although in the near future Azerbaijan would prefer to work with other producers to fill the pipeline, such as Turkmenistan, Iran, Israel, Cyprus and others. Politics could prevail, but economic interests will play a key role in the Italian government's final decision.

5.12. Greece's Role in Strengthening National and European Energy Security

The geopolitical significance of Greece was traditionally based more on its geostrategic value than its geo-economic status. In the middle of three continents and important seas (the Eastern Mediterranean, the Black Sea, and the Adriatic), its geophysical morphology in land and sea constitutes a geostrategic point of reference. Situated in the heart of the Mediterranean, Greece has been a perennial strategic trading post, enabling it to play a key role in the context of Mediterranean security. The United States (US), for example, considered Greece after World War II as part of the Southeast Mediterranean defense perimeter against Soviet expansionism. Due to the many islands, only a few waterways exist that allow big ships to go through, making navigation difficult but also easier to monitor. Mediterranean security is very important to Greece not only because of the historical legacy but also because a Mediterranean Sea free of tensions is imperative for the country's economic prosperity and security. For Greece, Eastern Mediterranean was, and still is, the great gateway from the East to the West, and there are significant shipping routes across the Eastern Mediterranean and the Aegean Sea. Crete is of strategic importance: located midway between mainland Greece and Egypt, it is surrounded by important sea routes that enhance the geopolitical importance of Greece. Substantial US aeronautical forces are also based on the island. As early as 1953 Joseph Roucek underlined that not only Crete, but also Cyprus, are of vital concern to the countries bordering the Eastern Mediterranean as well as to every great power, including the US (Roucek, 1953). Moreover, a large proportion of Greece's energy resources are imported through the Mediterranean basin via oil tankers from the Persian Gulf countries and elsewhere. Therefore, Greece's location makes it a natural bridge between the energy-rich Middle East and energy-consuming Europe. At the same time, Greeks are looking to transform the country into another potential hub —like Turkey— for bringing Eastern Mediterranean gas to European markets. On the other hand, Greece is a relatively small state. Its leadership understands that its political weight is limited to the regional rather than the international level. According to the New Geopolitics of the Eastern Mediterranean, the Greece was obliged to maintain its influence in the Eastern Mediterranean and the Middle East as well as in the Balkans

Greece has three intertwined grand strategy goals. The one is to act as an anchor of stability in its broader neighborhood. The second is the peaceful resolution of open disputes with neighbors, primarily Turkey, within the framework of international law and treaties. The third is to retain a relative balance of power with, and leverage its own position vis-à-vis, Turkey, with an eye to thwarting any unilateral Turkish moves that could threaten Greece's sovereign rights. In this context, Greek foreign policy is solidly tied to and embedded in US-led NATO. Nevertheless, NATO's lack of security guarantees to Greece vis-à-vis fellow NATO member Turkey, and perceptions of constant Russo Turkish friction, reinforce Greece's drive to maintain amiable relations, open channels of political communication and alliance rhetoric, and significant military trade ties with Russia. Greece's membership in the EU provides additional channels for Greek foreign policy and diplomacy and implicit guarantees of solidarity in case of Turkish aggressiveness. Becoming an energy hub may provide further guarantees to Greece's security and territorial integrity. Greek proclivity to support the EU gas diversification agenda is dictated by Greece's grand strategy. First, entering into the pipeline map and becoming an essential transit country will amplify Greek geopolitical advantage. Second, transit status provides an extra negotiating card for the country, which can be used in negotiations through issue linkages. However, these considerations remain vague and have yet to be translated into tangible outcomes in the service of (any of) Greece's three principal foreign policy goals. The above notwithstanding, underneath Greece's gas diplomacy includes two contradictions. First, the energy security consideration is dominated by the stark divide and incoherence between a Russia-centric and an anti-Russian (translated in policy terms as US-friendly) perspective. The latter sees dependence on Russian gas as unsafe and advocates a reduction of gas imports from, and no further infrastructure deals with, Russia. The former prioritizes good relations with Russia to ensure stable gas flows and enhance energy security. This feeds into an ambiguous gas diplomacy. On the one hand, Greece lies in the frontline of diversification projects, such as the Interconnector, TAP and EastMed pipelines and LNG receiving terminals, which serve US geopolitical interests. At the same time, Greece endorsed and joined the Russia-sponsored South Stream project and, subsequently, the IGI/Poseidon project that is designed to link up to Turkish Stream, both of which would perpetuate Greek and emphasize EU dependence on Russian gas. At a regulatory level in the EU context, Greece has backed a conciliatory position that would allow Gazprom to maintain its EU market share, rather than sided with those members pushing for effective limitation of Gazprom's share in the EU market. As a result, Greece seems to oscillate between policies related to both Russian and US energy interests. This, contrasts with its more refined foreign policy approach and may either compromise Greece's ability to stick to its foreign policy priorities or undermine its energy security. Secondly, Greece aims to advance its own geopolitical standing not only in vague, absolute terms, but also in relative terms, vis-à-vis Turkey. This competition is explicit, also taking into account Turkey's declared goal to evolve into an energy hub, and is played out across two antithetical and mutually exclusive transport projects: South Stream vs. Turkish Stream, and the EastMed pipeline vs. a pipeline shipping Israeli gas to Turkey (the TAP project advances both countries' interests in a non-conflictive way as it is transit-only for Greece). In the latter antithetical pair, options via Turkey have been muted for geopolitical reasons, but the EastMed option still comes up against significant obstacles. These include complicated geopolitics, low gas prices to finance new infrastructure, and difficulties in increasing regional supply enough to justify the cost of expensive pipelines (Prontera, 2017). The prospects hence remain uninviting for any pipeline project carrying East Mediterranean gas, let alone for both of them. Contrary to this lose-lose case, in the former (antithetical pair) Turkish Stream has won over South Stream; in case Greece aspires to becoming a transit state as well, this will be done in tandem with, rather than at the cost of Turkey's emerging gas hub status. This is an asymmetrical win-win situation as Turkey's emerging gas hub status is superior to Greece's transit status, and conveys more advantage to Turkey than to Greece, especially if one considers the energy security dimension analysed above. Hence, all in all, Greek endeavours to reap geopolitical benefits through energy remain inconclusive at best, counter-productive at worst (Proedrou 2019).

To sum up, Greece, according to its position, links Asia with Europe and Middle East through the sea, so it automatically becomes a suitable and convenient choice for transit state; there is not high risk of terrorism, there is no war or other significant conflicts inside the country or with the Union, no piracy, no strong dependence on one leader and finally, natural phenomena are not so strong to threat to infrastructures with a disaster. Consequently, there would be assured security in the supply chain at great extent. Greece can hold a key part in the transportation of energy from East to West through pipeline projects, electricity grid interconnectivity and alternative means of ensuring Security of Supply through Mediterranean offshore reserves (e.g., LNG terminals).

CONCLUSION

This book is an introductory approach, in the wider theoretical considerations on scientific research and study, of issues relating to energy resources and energy security and, at the same time, it is a case study on Eastern Mediterranean energy security. Its aim is to initially study, investigate and correlate certain basic terms and current issues: energy resources, energy networks, energy security.

The concept of energy security is a multidimensional and complex concept defined by various interpretations and includes areas such as the environment, the economy, security and geopolitics. It has been defined differently, for example, by the IEA and differently by the European Commission. Nevertheless, a common point of reference is the focus on the issue of ensuring the availability of energy on demand. For many, the issue of energy security is now perceived as something more important than the issue of climate change due to the immediacy of the effects that will occur if the latter is not addressed properly. In order to understand the range of issues that interact with energy security, one should emphasize the relevance of climate change and of the subsequent choice of the means used to produce and distribute energy. A second issue is the volatility of energy markets, with the most recent example being the recent drop in oil prices. In addition, it turns out that political developments have a catalytic effect on the energy sector. A typical example is the recent decision by

Saudi Arabia not to intervene in falling oil prices and to continue large-scale production, which directly affects the energy security of many countries and the oil market in general.

With such an unstable international environment at the political and geopolitical level and with the turmoil that exists, the concept of energy security is at the top of the agenda, especially for the EU. Attempts are being made to consolidate the energy markets of the Member States and to design a common foreign policy towards third countries, especially regarding the energy sector. The volatile energy landscape, in which new powers such as China and India are growing rapidly, the very important role that Russia is playing and the projected increase in energy demands in the coming years, have led to conflict. Searching for and controlling wealth-producing resources makes it easy to see that the issue of energy demand is a matter of immediate concern and capable of changing the global landscape.

More specifically, the discovery and widespread use of petroleum as fuel in the modern world as well as the transition of societies to hydrocarbons (strategic resources) have had multiple effects. Historically, energy security has been mainly associated with oil supply and the wider Eastern Mediterranean region. However, the increasing complexity of the energy systems requires systematic and thorough uptake as well as a deep understanding of a wide spectrum of important elements. Therefore, an analysis of the oil supply security does no longer provides enough ground for understanding the energy security condition of a State. Primarily, the vast interests in the arena of the international centres of power should be investigated and studied given that energy, on account of its great importance, is the bone of contention among the International Powers: the exploitation of the wealth of poor countries and the increased economic disparities are still prevalent, highlighting, among others, that energy resources are factors instrumental in maintaining power and supremacy on the planet.

Ankara's partnership with Moscow and Tehran, and its simultaneous withdrawal from the West, are drastically changing geopolitical facts and the distribution of power in the Middle East. However, it cannot be ruled out that Turkey's change of course will be temporary, since, despite its tensions with the United States and some European countries, it has not severed its ties with the West. The further Turkey moves away from the West, the more Greece's geopolitical position is upgraded.

Greece and Turkey are linked by mutual interests, due to the route of the "Southern Energy Corridor" that passes through the territory of our neighbor country before entering ours. The same will happen if the TurkStream extension is implemented within the Greek territory, before it ends up in European countries.

The fact that Turkey has been left out of the "Forum for Natural Gas in the Eastern Mediterranean" is a clear indicator of its further political isolation in the region. Such a development reinforces the belief that it will eventually stay out of the countries of transit of natural gas in the Southeastern Mediterranean to European markets. In this context and given that Ankara feels limited by the Greek and Cypriot maritime zone, it appears that it will increase its aggression in order to establish its opposition to the emerging geopolitical environment. As a result, it will continue to claim part of the region's gas production and transit in order to achieve its further geopolitical upgrade.

It is also natural that there will be increased pressure by the international community, to resume negotiations aimed at the resolution of the Cyprus issue. Resolving this thorny and chronic issue will help secure, to some extent, the critical energy region of the Southeastern Mediterranean from hotbeds of instability and conflict.

Egypt's potential pursuit of a strategic relationship with the EU, with Greece and Cyprus as intermediaries and connecting links, is geopolitically upgrading our country. Increasing hydrocarbon production from the United States enables them not only to meet their high energy requirements, but also to generate a significant surplus for exports. Therefore, in addition to the gradual withdrawal of the United States from energy imports, it is possible to gradually reduce the energy dependence of the old continent on the Middle East and Russia, and at the same time increase the diversity of suppliers and routes.

Although Moscow has achieved steady growth in trade agreements, military successes and open diplomacy with all sides, the current situation carries the risk of being temporary. Moscow's interaction with the Middle East can be described as a transactional one. And when interests clash, the agreements that are conflicting to the ones of the United States - which has been considered the guarantor of the security of some countries in the region for many years - will be affected.

Logically, Russia's relationship with the Middle East is likely to deteriorate should the geopolitical or economic interests so require. To build a long-term relationship, Moscow needs to institutionalize its cooperation and establish a resilient level of interaction. The agreement to reduce oil production may be an indication of the long-term strategies that can be implemented in a predetermined framework of cooperation.

Although the United States remains the strongest player in the region, it no longer has the footholds of the past, especially in the sub-system of the Gulf states. Russia is becoming a geopolitical and military partner for the regimes of the countries that oppose Washington, having regained its foothold in Syria and seeking to gain a foothold in Libya and Yemen. Its constant goal is to create the conditions for the promotion of its power in the region.

Moscow's influence in the Middle East is largely due to its economic ability to maintain military forces in Syria and generally to have the resources needed to gain a foothold in the Middle East. It is doubtful whether it will be able to continue to do so in the long run, based on the economic, military and demographic dimensions that characterize it, even in relation to the United States.

In the geopolitical environment of the Middle East, there is a multiculturalism that is mainly next to the differentiation of some countries such as Turkey, Egypt and Saudi Arabia from the close confrontation with the United States. The United States, which at least for now seems unwilling to get involved, is characterized by an unorganized and long-term strategy in the region. In this way, regional and super-systemic actors are given the opportunity to increase their influence in the Middle East. In fact, Russia seems to be evolving into a regulatory force, while Iran and

Turkey are also trying to increase their influence. Will this trend continue to exist, in the medium and long term, or will there be a dynamic intervention by the United States that will change the data?

In multiculturalism, at the level of national actors, which characterizes the geopolitical landscape of the Middle East, the supranational actors are additionally added. These are organizations, coalitions and consortiums of economic interests that in turn claim part of the power of traditional state entities, mainly at the economic level but also at the level of political influence.

The future of the Iranian economy and Iran in general is uncertain because of the continuing pressure from US sanctions. Due to the dire economic situation in the country, China and Russia are likely to increase their presence in Iran, where they will replace European companies, which may leave due to increased uncertainty. After such a development, the question that arises is whether or not the EU will have reasons to stay in the JCPOA.

TAP may be delayed due to the failure to secure the necessary gas quantities from Azerbaijan, its changing architecture, and the Italian reaction. The above obstacles are considered to be incapable of making it irrevocably inactive. From a geopolitical point of view, the EastMed pipeline is an important pipeline for the West, but without being an antidote to Russian gas. In the longer term, exploitation of other deposits in the Southeastern Mediterranean, including possibly those in Greece and Egypt, could further strengthen Europe's energy security.

Russian gas supplies to Southeast Europe remain uncertain after 2020, as Russian pipelines through Ukraine are scheduled to be shut down, and TurkStream's "exit" to European markets has not been finalized. Greece now has the potential to contribute to the diversification of energy routes and power supplies, as well as the consolidation of the European energy market. Our country's contribution to strengthening European energy security, at the regional level, could be particularly critical due to the continuing instability in the wider region.

All the findings of the study demonstrate certain dimensions of a reality, which are expected to be dynamic creating risk but also future opportunities. The main finding of the study points not only to the strong correlation between energy power and economics and politics but also to the increased dependence of the latter two on the former, thus creating a hotbed of tensions, rivalries, and conflicts every time the existing relationship between economy and energy are destabilized.

The shifting to energy sources, which are safe and eco-friendly, is a one-way process, the only means of survival and safe development for humanity. Generating energy based on economic, safe, cost-efficient and renewable criteria is the current ecological and technological challenge provided that, firstly, the vast interests in the energy sector are dealt with successfully.

In summary, the Eastern Mediterranean has been a region with distinctive characteristics forming the basis for collaboration and dispute for them. Furthermore, the wider Eastern Mediterranean region will continue to play a vital role in energy security, which has been termed as the greatest Geostrategic "game" of the next fifty years. The importance of 'regional interdependence' should be further emphasized so as to reveal the conditions and benefits of security and cooperation in the region and the role of the EU in the region must be further explained and clarified especially as to how the EU helps to respond to the current challenges or exacerbate crises. Innovative methods and approaches should be developed in order to enhance the EU's visibility and impact in the region. In these days, issues like energy and migration have emerged as the main drivers of both cooperation and contention. However, regional cooperation for the management of these issues and relative challenges seem to be the sole solution for effective and sustainable development of the area.

About the Author

Michail Chalaris, MA, PhD
Department of Chemistry, International Hellenic University,
Kavala, Greece

Dr. Michail Chalaris received PhD in Chemistry from the Department of Chemistry of the School of Sciences, National and Kapodistrian University of Athens and BSc degree in Chemistry from the same Department. Furthermore, he received 2nd BSc from Fire Officers Faculty of Hellenic Fire Academy and MA in Applied Strategy and International Security from University of Plymouth. Michail is a graduate of Hellenic National Defence College and of School of Further Education & Professional Training of Hellenic Fire Academy

Retd Major General (HFCs) Michail Chalaris is a 25-year veteran of thc fire service. He brings a wide variety of experiences, a high level of energy, and a sincere passion for his instruction and service. His expertise addresses the intersection of first-hand operational experience in all-hazards mitigation, crisis management policy, as well as and functional elements of communications. This includes information communications technology crisis management, incident command systems for large events (e.g., Olympics) and catastrophes, incident action plans, as well as the development and execution of train-the-trainer programs. Michalis's strengths also include all-risk and all-incident leadership and team building in a crisis environment requiring rapid set up. In particular, he excels in training key stakeholders on how to manage and command, as well as the use of the virtual space. He is Certified Safety Officer. Furthermore, he is (a) Technical Expert of European Commission/Humanitarian Aid & Civil Protection, (b) Member/Expert of Hellenic CBRN Incidents Support Team and (c) Member of Coordinative Committee of Hellenic Network of Green Chemistry.

He is Assistant Professor at Department of Chemistry of International Hellenic University (2020 -) and Adjunct Professor at Hellenic Fire Academy (1999 – until now), Military Nursing Academy (SAN) (2017 – until now) and Professor in MSc "Analysis and Management of Manmade and Natural Disasters" in the course Technological disaster and environmental emergencies (2015 – until now), in MSc Oil and Gas Technology in the course Hazard Identification & Risk Management (2018 – until now). He is also tutor at the National School of Public Administration and Local Government (ESDDA) (2005 – until now) and a certified trainer by National Organization for the Certification of Qualifications & Vocational Guidance (EOPPEP) (1996 – until now). He has been President of Association of Greek Chemists (2004-2005) and Dean of the Fire Officers Faculty of Hellenic Fire Academy (2015 -2016). From 2009 to 2012, he served as State Executive Secretary of Labour Inspectorate in Ministry of Labour, Social Security and Welfare. Finally, he was (i) member of Working Group Education and Training (2013 - 2018) and (ii) of Hazmat Commission (2018 – 2019) of CTIF, (iii) Chairman of the National Committee for Examination of Proposals for the participation of the Hellenic Fire Corps in research programs (2014 -2018), (iv) member of national council of policy and control of foods. under the ministry of development (2006-2010), (v) member of national council of health and safety at work. under the ministry of labor and social affairs (1999-2010) and (vi) national delegate in the committee physical and biophysical chemistry of IUPAC (2005-2008).

He participated as leader of behalf of Association of officers and sub-officers with university degrees of Hellenic Fire Corps (GR: Ε.Π.Α.Υ.Π.Σ.) (EN: E.P.A.Y.P.S.) in two projects

1. H2020-SU-SEC-2019 Type of action: RIA, Proposal number: 882897, Proposal acronym: Search and Rescue, Duration (months): 36, Proposal title: Search and Rescue: Emerging technologies for the Early location of Entrapped victims under Collapsed Structures and Advanced Wearables for risk assessment and First Responders Safety in SAR operations, Activity: DRS02.
2. UCPM-2020-KN-A, Proposal number: 101017819, Proposal acronym: RESISTANT, Proposal title: Training and Knowledge Sharing Platform For First Responders and Educational Tools for students' and citizens' awareness and preparedness against Natural and Manmade Disasters and Risks

Furthermore, is the contact person on behalf of EPAYPS (2016 – until now) and he has been working closely through signing MoU with National Yunlin University of Science and Technology (YunTech), Central Police University, National Fire Agency and New Taipei City Government Fire Department from Taiwan and the National Volunteer Fire Council (NVFC) of USA. The purpose of these MoUs is to continue to develop and expand a framework of cooperation between the organizations to develop mutually beneficial programs, projects and activities in areas of mutual interest. Moreover, participate as full parties in a Consortium for Research, Education, Advanced Technologies & Environment (CREATE) which is an International Consortium for Research and Innovation.

He works closely with the Humanitarian and Development Research Initiative (HADRI), at Western Sydney University (Australia) and UNESCO Chair on Conservation and Ecotourism of Riparian and Deltaic Ecosystems, International Hellenic University (Greece) for the following proposals (a) Hydropower for Disaster Resilience Applications (HYDRA) & (b) Renewable Energy Generated by the Impacts of Natural and

Accidental Disasters. Moreover, he stars a research for the managing intermodal transportation of hazardous materials accident.

Finally, he was member of the Hellenic EU Presidency's Civil Protection Team (2014), taking part in both the organisation and the execution of the Tabletop and Full-Scale Exercises (TTX and FSE) PROMETHEUS 2014 and EVITA 2014. Dr. Chalaris is a program committee member of several international conferences. He has organized/coordinated or participated as a keynote and invited lecturer/speaker on several advanced training courses and advance research workshops. He also managed/participated in several multiyear projects related to the area of his expertise.

Editor in Chief of the book entitled Novel Approaches in Risk, Crisis and Disaster Management (Nova Science Publishers, Inc. ISBN: 978-1-53613-239-7, 2018). Reviewer of international journals and member of their Editorial Boards: (a) SCIREA Journal of Safety Science and Technology, (b) SCIREA Journal of the Environment & (c) CBRNE-Terrorism Newsletter.

His research interests among other fields focus on: (a) Physical Chemistry, Theoretical Chemistry, Computational Chemistry, Computer simulation, Molecular Dynamics, Molecular Dynamics Simulation, Molecular Modelling, Monte Carlo Simulation, Computational Materials Science, Hydrogen Bonding, Statistical Mechanics, Material Modelling, Statistical Thermodynamics, (b) International Security, Diplomacy, Terrorism, WMD, CBRN(E), Hazardous Materials (c) Risk-Safety and the Environment, Risk Assessment for Environmental Emergencies, Technological accidents, Occupational Health and Safety, Chemical Safety, Fire Safety, Hazard Assessment, Solvents, Chemical Hazard Release, (d) Labour Law, undeclared work posted workers and (e) Vocational Education.

Up to today he have been granted scholarships from: A.S. Onasis (1-10-1989 until 31-6-1990), the Institute of Electronic Structure and Laser of ITE (1997), the ESF (European Science Foundation) 1998 and the EMLG (European Molecular Liquids Group) 1997, 1999, 2000 and 2001 for the presentation of his research work.

Publications from the Last 3 Years:

1. Spyros Schismenos, Garry J Stevens, Dimitrios Emmanouloudis, Nichole Georgeou, Surendra Shrestha and Michail Chalaris (2020). Humanitarian Engineering at the Sustainability-Development Nexus: Mapping Vulnerability and Capability Factors for Communities at Risk of Water-Based Disasters. *Sustainability Science.*
2. Michail Chalaris, Vangelis Pipitsoulis, Ray Ingblum (2020). *Long-Distance Voice Messaging Systems (Clear Voice) During The Time Of Natural And Manmade Disaster According To Health And Safety Regulations.*
3. SafeGreece *2020 on-line Proceedings – 7*th *International Conference on Civil Protection & New Technologies* 14-16 October, pp 205-208, [https://safegreece.org/safegreece2020/images/docs/safegreece2020_proceedings.pdf] Publisher: SafeGreece ISSN 2654-1823
4. Michail Chalaris, Vangelis Pipitsoulis, Itamar Banayan (2020). *Novel Solutions For Command And Control Network For Disaster Management.*
5. SafeGreece *2020 on-line Proceedings – 7th International Conference on Civil Protection & New Technologies* 14-16 October, pp 189-192, [https://safegreece.org/safegreece2020/images/docs/safegreece2020_proceedings.pdf] Publisher: SafeGreece ISSN 2654-1823.
6. Michail Chalaris, Conrad Bielski, John Alexander, Vangelis Pipitsoulis, Ioannis Kassaras (2020). I-React: Improving Resilience To Emergencies Through Advanced Cyber Technologies. SafeGreece *2020 on-line Proceedings – 7*th *International Conference on Civil Protection & New Technologies* 14-16 October, pp 165-168, [https://safegreece.org/safegreece2020/images/docs/safegreece2020_proceedings.pdf] Publisher: SafeGreece ISSN 2654-1823.

7. Nikos Tsalas, Giorgos Katsouras, Michalis Chalaris, Stelios Samios, Efthymios Lytras, Kostas Papadopoulos, Nikos Kouris, Simos Malamis, Costas Noutsopoulos, Danos Mamais, Andreas Andreadakis (2020). Novel Systems For Detection And Monitoring Of Dangerous Substances In Water Environmental Sensitive Areas. SafeGreece *2020 on-line Proceedings – 7th International Conference on Civil Protection & New Technologies* 14-16 October, pp 141-144, [https://safegreece.org/safegreece 2020/images/docs/safegreece2020_proceedings.pdf] Publisher: SafeGreece ISSN 2654-1823.
8. Chih-Long Pan, Michail Chalaris, Jet-Chau Wen (2020). An Innovative Emergency Medical Regulation Model In Natural And Manmade Disasters. SafeGreece 2020 on-line Proceedings – 7th International Conference on Civil Protection & New Technologies 14-16 October, pp 27-30, [https://safegreece.org/safegreece2020/ images/docs/safegreece2020_proceedings.pdf] Publisher: SafeGreece ISSN 2654-1823.
9. I. Kassaras, V. Pipitsoulis, P. Papadimitriou, N. Voulgaris, V. Kouskouna, G. Kaviris, V. Kapetanidis, A. Karakonstantis, E. Lekkas, S. Lalechos, P. Bernard, R. Bossu, L. Fallou, F. Finazzi, C. Bielski, M. Chalaris, A. Mitropoulos, R. Kosheleva, P. Kinanlis, D. Kazantzidou-Firtinidou, L. Stupina (2020). I-REACT RSRA: Development of a Rapid Seismic Risk Assessment Tool That Integrates Instrumental, Crowdsourcing, Social Media Data and Empirical Vulnerability. SafeGreece 2020 on-line *Proceedings – 7th International Conference on Civil Protection & New Technologies* 14-16 October, pp 169-172, [https://safegreece.org/ safegreece2020/images/docs/safegreece2020_proceedings.pdf] Publisher: SafeGreece ISSN 2654-1823.
10. Christos Seretidis, Dimitrios Billios and Michail Chalaris (2020). *Refugee Crisis: Global Perspectives, Challenges and Issues* (Editor Paul Leclair), Chapter 1. How Technology Is Helping to Face the Illegal Immigration Crisis and to Modernize Border Security, Nova Science Publishers, Inc. ISBN: 978-1-53618-522-5.

11. Michail Chalaris (2020) *Refugee Crisis: Global Perspectives, Challenges and Issues* (Editor Paul Leclair), Chapter 2. Illegal Immigration: Is It a Threat to National Security. The Greek Experience, Nova Science Publishers, Inc., ISBN: 978-1-53618-522-5
12. Spyros Schismenos, Garry John Stevens, Dimitrios Emmanouloudis, Nichole Georgeou, Surendra Shrestha, and Michail Chalaris (2020). Humanitarian engineering and vulnerable communities: hydropower applications in localised flood response and sustainable development. *International Journal of Sustainable Energy*, Vol. 39:10, pp 941-950 https://doi.org/10.1080/14786451.2020.1779274
13. Nikolaos Stasinopoulos & Michail Chalaris. (2020). A Disaster Resilience Approach Applying in a Smart City Concept. *Case Studies Journal*, ISSN (2305-509X), pp. 64 -66, Volume 9, Issue 3–March-2020.
14. Meliadis Miltiadis, Sipitanou Afroditi & Chalaris Michail. (2020). Study of the Problems Associated with Producing a Satisfactory Definition of Terrorism and Analysis of the Main Characteristics of Terrorism. *Case Studies Journal*, ISSN (2305-509X), pp. 55 - 63, Volume 9, Issue 3–March-2020.
15. Sofia Karma, Spyros Schismenos, Dimitrios Emmanouloudis & Michail Chalaris. (2019). *Challenges and Lessons Learned from past major Environmental Disasters due to Technological or Wildland Urban Interface Fire Incidents.* Full Article 30 p - Contributing Paper to GAR 2019 (ISBN/ISSN/DOI978-92-1-004180-5) https://www.undrr.org/publication/challenges-and-lessons-learned-past-major-environmental-disasters-due-technological-or.
16. Michail Chalaris. (2019). Analysis of Bleve Scenario in the LPG Tank: A Case Study at the Kamena Vourla Incident. SafeCorfu *2019 Proceedings – 6th International Conference on Civil Protection & New Technologies*, Volume: Page I 78- 81 – ISSN 2654-1823, SafeGreece.

17. Nikolaos Stasinopoulos & Michail Chalaris. (2019). Smart Cities: Their Role in Protection from Natural and Man-Made Disasters. SafeCorfu *2019 Proceedings – 6th International Conference on Civil Protection & New Technologies*, Volume: Page II 25- 27 – ISSN 2654-1823, SafeGreece.
18. Michail Chalaris. (2019). Illegal Immigration as a National Security Problem. The Greek Experience. SafeCorfu *2019 Proceedings – 6th International Conference on Civil Protection & New Technologies*, Volume: Page II 17- 27 – ISSN 2654-1823, SafeGreece.
19. Kontos Aristeidis & Chalaris Michail. (2019). Assessing the Skilful Use of Diplomacy as a Useful Tool to Increase a State's Power and Influence. *Conference Proceedings of International Scientific Conference "Challenges and Strategies in Public Order and Safety"* 8th Edition, pp. 70 – 76, ISSN 2559-3277, (December 6, 2019). Published by Sitech Publishing House Craiova, Romania, 2019 Editors: Prof. univ. dr. Ioan Mara & Lect. univ. dr. Băncilă Adrian.
20. Gkyzi Eleni & Chalaris Michail. (2019). An Analysis of Terrorist Threat in Cyber-Space as an Emerging Threat. *Conference Proceedings of International Scientific Conference "Challenges and Strategies in Public Order and Safety"* 8th Edition, pp. 184 - 191, ISSN 2559-3277, (December 6, 2019). Published by Sitech Publishing House Craiova, Romania, 2019 Editors: Prof. univ. dr. Ioan Mara & Lect. univ. dr. Băncilă Adrian.
21. Natsis Dimitrios, Kokkalis Ioannis, Lampi Evgenia & Chalaris Michail. (2019). Study Of Conditions That Are Necessary For A Terrorist Organization Or Group To Survive. *Conference Proceedings of International Scientific Conference "Challenges and Strategies in Public Order and Safety"* 8th Edition, pp. 192 - 198, ISSN 2559-3277, (December 6, 2019). Published by Sitech Publishing House Craiova, Romania, 2019 Editors: Prof. univ. dr. Ioan Mara & Lect. univ. dr. Băncilă Adrian.

22. Meliadis Miltiadis, Sipitanou Afroditi & Chalaris Michail. (2019). Study And Evaluation Of The Ways In Which Terrorist Organizations Seek To Exploit The Media. *Conference Proceedings of International Scientific Conference "Challenges and Strategies in Public Order and Safety"* 8th Edition, pp. 199 - 207, ISSN 2559-3277, (December 6, 2019). Published by Sitech Publishing House Craiova, Romania, 2019 Editors: Prof. univ. dr. Ioan Mara & Lect. univ. dr. Băncilă Adrian.
23. Lampi Evgenia & Chalaris Michail. (2019). Terrorist Use Of A Chemical, Biological Or Radioactive Weapon Capable Of Causing Widespread Death And Destruction: How Serious Is The Threat? *Conference Proceedings of International Scientific Conference "Challenges and Strategies in Public Order and Safety*" 8th Edition, pp. 389 – 398, ISSN 2559-3277, (December 6, 2019). Published by Sitech Publishing House Craiova, Romania, 2019 Editors: Prof. univ. dr. Ioan Mara & Lect. univ. dr. Băncilă Adrian.
24. Spyros Schismenos, Garry Stevens, Dimitrios Emmanouloudis, Nichole Georgeou, Michail Chalaris & Antoine Smith. (2019). Local Disaster Vulnerability Analysis: An Approach To Identify Communities. UN MGCY Youth Science Policy Interface Publication – 2nd Special Edition: *Disaster Risk Reduction: Moving Forward, Thinking Ahead.* pp. 49-54, August 2019. https://www.preventionweb.net/files/68036_ 68036drrunmgcy2019final.pdf.
25. Antoine Smith, Spyros Schismenos, Garry Stevens, Leonard Hutton, Michail Chalaris & Dimitrios Emmanouloudis. (2019). Understanding Large-Scale Fire Events: Megafires In Attica, Greece And California, USA. UN MGCY Youth Science Policy Interface Publication – 2nd Special Edition: *Disaster Risk Reduction: Moving Forward*, *Thinking Ahead*., pp. 29-34, August 2019. https://www.preventionweb.net/files/68036_68036drrun mgcy2019final.pdf.
26. Georgios Romosios, Georgios Ntzaferis & Michail Chalaris. (2018). The Boilover Phenomenon: Requirements For Its

Development, Calculation Of Its Manifestation Time And Firefighting Tactics. Safekozani 2018 *Proceedings – 5th International Conference on Civil Protection & New Technology*, pp. 134 - 142 - ISSN 2654-1823, SafeGreece.

27. Sofia Karma, Milt Statheropoulos, Olga Kakaliagou, Ioannis Boukis, Evangelia Pelli & Michail Chalaris. (2018). Disaster Preparedness & Response For Vulnerable Groups Of Population: Evacuation Planning Of Critical Infrastructures In Case Of An Earthquake Or A Fire For People With Disabilities. Safekozani *2018 Proceedings – 5th International Conference on Civil Protection & New Technology*, pp. 314 – 318 - ISSN 2654-1823, SafeGreece.
28. Lucas Rocha, Leonardo Dias, Pedro Peregrini, Pedro Lima, Spyros Schismenos, Dimitrios Emmanouloudis & Michail Chalaris. (2018). A Database Development in Nova Friburgo Town for Analysis and Complementation of the Mapping Methodology Proposed in the Gides's Project. Safekozani 2018 *Proceedings – 5th International Conference on Civil Protection & New Technology*, pp. 334- 318 - ISSN 2654-1823, SafeGreece.
29. Spyros Schismenos, Michail Chalaris, Dimitrios Emmanouloudis & Nikolaos Katopodes. (2018). Renewable Energy and Drones In Search and Rescue: Automated Network For Air-Sea Actions. Safekozani *2018 Proceedings – 5th International Conference on Civil Protection & New Technology*, pp. 358-365 - ISSN 2654-1823, SafeGreece.
30. Spyros Schismenos, Michail Chalaris, Dimitrios Emmanouloudis, Nikolaos Katopodes & Garry Stevens. (2018). Renewable Energy Generated By the Impacts of Natural and Accidental Water-Based Disasters. Safekozani *2018 Proceedings – 5th International Conference on Civil Protection & New Technology*, pp. 391 - 397 - ISSN 2654-1823, SafeGreece.
31. Schismenos, Karma, Chalaris. (2018). *Novel Approaches in Risk, Crisis and Disaster Management*, Editor in Chief Michail Chalaris Chapter 5. Large-Scale Fire Incidents in Recycling Plants: Lessons

Learned from Two Indicative Case Studies and Future Needs. Nova Science Publishers, Inc. ISBN: 978-1-53613-239-7.

32. Schismenos, Chalaris, Gazouli. (2018). *Novel Approaches in Risk, Crisis and Disaster Management*, Editor in Chief Michail Chalaris Chapter 6. Nanotechnology Applications for Biological Threat Detection. Nova Science Publishers, Inc., ISBN: 978-1-53613-239-7.
33. Patitsa, Chalaris. (2018). *Novel Approaches in Risk, Crisis and Disaster Management*, Editor in Chief Michail Chalaris. Chapter 13. Working with Refugees in a Disaster Crisis: The Development of Secondary Post-Traumatic Stress among Emergency Workers and First Responders. Nova Science Publishers, Inc. ISBN: 978-1-53613-239-7.
34. Michail Chalaris. (2017). *Nuclear Weapons: Global Programmes, Challenges and Security Implications*. Chapter 1. Preparedness and Response to Nuclear Crisis Synopsis of Essential International Obligations and Cooperation Mechanisms. Nova Science Publishers, Inc. ISBN: 978-1-53611-860-5.
35. Evacuation Planning of Critical Infrastructures in case of an Earthquake or a Fire for People with Disabilities. (2017). *European Center for Forest Fires*, ISBN: 978-618-83079-0-2 2016.
36. Michail CHALARIS. (2017). Preparedness and Response to Nuclear Crisis Synopsis of Essential International Obligations and Cooperation Mechanisms. *International Journal of Terrorism & Political Hot Spots*, ISSN: 1932-7889, Vol. *12*, Issue 1, p. 41-55.

References

Andrei, R. (2015). Energy Security in South-East Europe: Natural Resources as Causes of conflict or Building Stability. *Balkan Social Science Review*, (5), 63-87.

Arena, F., Laface, V., Malara, G., Romolo, A., Viviano, A., Fiamma, V., & Carillo, A. (2015). Wave climate analysis for the design of wave energy harvesters in the Mediterranean Sea. *Renewable energy*, *77*, 125-141.

Asvestopoulos, E. (2018). *Egypt's energy reserves, and the role of the EU on a potential Egypt-Israeli energy cooperation.*

Bagiorgas, H. S., G. Mihalakakou, S. Rehman and L. M. Al-Hadhrami, (2012). Offshore wind speed and wind power characteristics for ten locations in Aegean and Ionian Seas, *Journal of earth system science*, 121(4), 975-987.

Balodimos, S. (2016). *The energy resources of the Eastern Mediterranean.* The role of Turkey.

Bichler, S., & Nitzan, J. (2014). E*nergy conflicts and differential profits:* An update. Research Note, 1-5.

Bidder, S., & Nitzan, J. (1996). Putting the state in its place: US foreign policy and differential capital accumulation in Middle East 'energy conflicts.' *Review of International Political Economy*, 3(4), 608-661.

Borchert, H., & Forster, K. (2007). Energy Infrastructure Security: Time for a Networked Public-Private Governance Approach. Middle East *Economic Survey*, 49, 21.

Bornstein, R. (2018). *Eastern Mediterranean regional dynamics: conflicts and opportunities for conflict resolution support.*

Boudière, E., Maisondieu, C., Ardhuin, F., Accensi, M., Pineau-Guillou, L., & Lepesqueur, J. (2013). A suitable metocean hindcast database for the design of Marine energy converters. *International Journal of Marine Energy,* 3, e40-e52.

Bradshaw, M. (2009, September). *The Geopolitics of Global Energy Security*. Geography Compass, 1920-1937.

Bronson, R. (2014). The United States in the Middle East: Bound by Growing Energy Demand. *Middle East Policy*, 21(2), 34-39.

Cherp, A., & Jewell, J. (2014, October 31). The Concept of Energy Security: Beyond the four As'. *Energy Policy*, 415-421.

Chorev, S. (2018, February). Security and Energy in the Eastern Mediterranean: The Israeli Perspective. In *Maritime Security in the Eastern Mediterranea*n (pp. 47-62). Nomos Verlagsgesellschaft mbH & Co. KG.

Cordesman, A. H., & Al-Rodhan, K. R. (2006). *The changing dynamics of energy in the Middle Eas*t (Vol. 2). Greenwood Publishing Group.

Communication of the European Commission on short-term Resilience (COM(2014) 654 final). (2014, October 16). Retrieved from https://eur-lex.europa.eu/legal-content/EN/TXT/?uri=CELEX:52014 DC0654.

Communication of the European Commission to the Council and the European Parliament: The European Energy Security Strategy. (2014, May 28). Retrieved from https://eur-lex.europa.eu/legal- content/EN/ ALL/?uri=CELEX:52014DC0330&qid=1407855611566.

Council Conclusions on Energy Diplomacy. (2015, July 20). Retrieved from http://data.consilium.europa.eu/doc/document/ST-10995-2015-INIT/en/pdf.

Dagoumas, A., & Flouros, F. (2017). Energy policy formulation in Israel following its recent gas discoveries. *International Journal of Energy Economics and Policy*, 7(1), 19-30.

DGEnergy. (2018a). *New era in EU- Egypt energy cooperation* Retrieved from https://ec.europa.eu/info/news/new-era-eu-egypt-energy-cooperation-2018- apr-23_en

DGEnergy. (2018b). *Quarterly Report on European Gas Markets.* Retrieved from https://ec.europa.eu/energy/sites/ener/files/documents/quarterly_report_on_european_gas_markets_q4_2017_final_20180323.pdf.

Different Types of Oil and Gas Company. (2015). *Oil and Gas Stakeholders.* Retrieved from https://www.planete-energies.com/en/medias/close/different- types-oil-and-gas-company.

Dimou, A. (2017). Energy as an east Mediterranean opportunity and challenge. *National Security and the Future*, 18(1 - 2).

Diaconu, S., & Rusu, E. (2013). The environmental impact of a Wave Dragon array operating in the Black Sea. *The Scientific World Journal,* 2013.

Dogru, S., & Reginbogin, H. (2017). Rethinking East Mediterranean Security: Powers, Allies & International Law. *Touro L. Rev.,* 33, 827.

Dokos, Thanos (ed.) (2016). *The Eastern Mediterranean in 2020: Possible Scenarios and Policy Recommendations, Policy Paper, Eliamep and the Konrad Adenauer Stiftung*, 1-36.

Dunn, T. (2012). *Neofunctionalism and the European Union.* Retrieved from https://www.e-ir.info/2012/11/28/neo-functionalism-and-the-european-union/

Elkind, J. (2017). *United States energy sector collaboration with Israel.* Hampton Roads International Security Quarterly, 5.

2030 Energy Strategy. (2018). Retrieved from https://ec.europa.eu/energy/en/topics/energy-strategy-and-energy-union/2030-energy-strategy.

Energy Security Strategy. (2014). Retrieved from https://ec.europa.eu/energy/en/topics/energy-strategy-and-energy- union/energy-security-strategy.

Energy: a shaping factor for regional stability in the Eastern Mediterranean? (2017). EU Parliament.

Energy Union Package, *A Framework Strategy for a Resilient Energy Union with a Forward-Looking Climate Change Policy*, Communication from the Commission to the European Parliament, The Council, The European Economic and Social Committee, The Committee of the Regions and the European Investment Bank, COM(2015) 80 final, European Union, Brussels, 25/02/2015, 2-3.

Eran, O. (2018). *Could Natural Gas in the Eastern Mediterranean Spark the Third Lebanese War?* Retrieved from http://www.inss.org.il/ publication/natural-gas-mediterranean-spark-third-lebanese-war/? utm_source=activetrail&utm_medium=email&utm_campaign=INSS% 2 0Insight%201021+1022+1023.

EU. (2018). *Energy Security Strategy*. Retrieved from https://ec.europa.eu/ energy/en/topics/energy-strategy-and-energy- union/energy-security-strategy.

Euractiv. (2018). *EU warns Turkey over Cyprus ship incident.* Retrieved from https://www.euractiv.com/section/politics/news/eu-warns-turkey-over- cyprus-ship-incident/

EUR-Lex: Regulation (Eu) 2017/1938 of the European Parliament and of the Council. (2017, October 28). Retrieved from https://eur-lex.europa.eu/eli/reg/2017/1938/oj

European Commission Press Release: Early Estimates of CO2 Emissions from Energy Use. (2019, May 8). Retrieved from http://europa.eu/ rapid/press-release_STAT-19- 2448_en.htm

European Commission Press Release: The Energy Union: From Vision to Reality. (2019, April 9). Retrieved from http://europa.eu/rapid/press-release_IP-19-1876_en.htm

European Commission Pressrelease: *Gas stress test: Cooperation is key to cope with supply interruption*. (2014, October 16). Retrieved from https://ec.europa.eu/energy/sites/ener/files/documents/IP-14-1162_EN. pdf

European Commission: *Liquefied Natural Gas*. (n.d.). Retrieved from https://ec.europa.eu/energy/en/topics/oil-gas-and-coal/liquefied-natural-gas-lng European Commission: *Security of Gas Supply*. (n.d.). Retrieved from https://ec.europa.eu/energy/en/topics/energy-security/secure-gas-supplies.

European External Action Service: *EU Global Strategy*. (2016, June). Retrieved from http://eeas.europa.eu/archives/docs/top_stories/pdf/eugs_review_web.pdf.

Eurostat: Energy Production and Imports. (2018, September 18). Retrieved from https://ec.europa.eu/eurostat/statistics- explained/index.php/Energy_production_and_imports.

Eurostat: EU Imports of Energy Products - Recent Developments. (2019, April 9). Retrieved from https://ec.europa.eu/eurostat/statistics-explained/index.php/EU_imports_of_energy_products_-_recent_developments.

Eurostat: EU Imports of Energy Products- Recent Developments. (2018, November 19). Retrieved from https://ec.europa.eu/eurostat/statistics-explained/pdfscache/46126.pdf IV.

Eurostat: Greenhouse Gas Emission Statistics- Emission Inventories. (2018, June 25). Retrieved from https://ec.europa.eu/eurostat/statistics-explained/pdfscache/1180.pdf.

Eurostat: Renewable Energy Statistics. (2019, January). Retrieved from https://ec.europa.eu/eurostat/statistics-explained/pdfscache/7177.pdf Eurostat: Shedding Light on Energy in the EU; A Guided Tour of Energy Statistics. (2018). Retrieved from https://ec.europa.eu/eurostat/cache/infographs/energy/bloc-3a.html.

Eurostat. (2018). *EU imports of energy products - recent developments*. Retrieved from: http://ec.europa.eu/eurostat/statistics-explained/pdfscache/46126.pdf.

Fawcett, L. (2016). *International relations of the Middle East*. Oxford University Press.

Gallis, P. (2006). *NATO And Energy Security*. Library Of Congress Washington Dc Congressional Research Service.

Gas and oil supply routes. (2018). European Commission Retrieved from https://ec.europa.eu/energy/en/topics/imports-and-secure-supplies/gas-and- oil-supply-routes.

Giannakopoulos, Angelos (ed.) (2016). *Energy Cooperation and Security in the Eastern Mediterranean: A Seismic Shift towards Peace or Conflict?* Research Paper 8, The S. Daniel Abraham Center for International and Regional Studies, Tel Aviv University, 1-132.

Cohen, E. (2018). Development of Israel's natural gas resources: Political, security, and economic dimensions. *Resources Policy*.

Crellijé, A., & van der Linde, C. (2006). European Supply Security and Geopolitics: A European Perspective. *Energy Policy*, 532-543.

Griffiths, S. (2017). A review and assessment of energy policy in the Middlc East and North Africa region. *Energy Policy*, 102, 249-269.

Halliday, F. (2005). *The Middle East in international relations: power, politics and ideology* (Vol. 4). Cambridge University Press.

Hinnebusch, R. (2015). *The international politics of the Middle East.* Manchester University Press.

Holland, R. A., Scott, K. A., Flörke, M., Brown, G., Ewers, R. M., Farmer, E., & Barrett, J. (2015). Global impacts of energy demand on the freshwater resources of nations. *Proceedings of the National Academy of Sciences*, 112(48), E6707-E6716.

Horden, P., & Purcell, N. (2019). *The Boundless Sea: Writing Mediterranean History*. Routledge.

Hughes, L. (2009). The four 'Rs' of Energy Security. *Energy Policy*, 2459-2461.

IEA. (2015). *Qatar: Natural Gas for 2015*. Retrieved from https://www.iea.org/statistics/statisticssearch/report/?country=Qatar&product=naturalgas.

IEA. (2018). *What is energy security?* Retrieved from https://www.iea.org/topics/energysecurity/whatisenergysecurity/.

Israel, Cyprus, and Greece push East Med gas pipeline to Europe. (2018). Retrieved from https://www.timesofisrael.com/israel-cyprus-and-greece-push-east-med- gas-pipeline-to-europe/.

İşeri, E. (2019). 14 Turkey's entangled (energy) security concerns and the Cyprus question in the Eastern Mediterranean. *Greece and Turkey in Conflict and Cooperation*, 257.

Jensen, A., Collins, K., & Lockwood, A. P. (Eds.). (2012). *Artificial reefs in European seas.* Springer Science & Business Media.

Kaplan, R.D. (2014). The geopolitics of energy. *Forbes* in https://www.forbes.com/sites/stratfor/2014/04/04/the-geopolitics-of-energy/#12db622d3b39.

Kim, Y., & Blank, S. (2015). US shale revolution and Russia: shifting geopolitics of energy in Europe and Asia. *Asia Europe Journal*, 13(1), 95-112.

Kiriyama, E., & Kajikawa, Y. (2014). A multilayered analysis of energy security research and the energy supply process. *Applied Energy*, 123, 415-423.

Klare, M. T. (2006). The Futile Pursuits of Energy Security by Military Force. *Brown J. World Aff.,* 13, 139.

Klare, M. T. (2015). Hard power, soft power, and energy power. *Foreign Affairs*, 3.

Kontos, M. (2019). Cyprus and the great powers an evaluation of Russian-cypriot relations in the new eastern Mediterranean environment. *Право и управление.* XXI век, 15(4), 17-27.

Koppa, M. (2017). *The Common Defense and Security Policy: History, Institutions and Strategies*, Patakis, Athens.

Leverett, F., & Bader, J. (2005). Managing China-US energy competition in the Middle East. *The Washington Quarterly*, 29(1), 187-201.

Litsas S. N. and Tziampiris A. (eds.). (2015). *The Eastern Mediterranean in Transition: Multipolarity, Politics and Power*, Ashgate, Surrey, Farnham.

López, I., Andreu, J., Ceballos, S., de Alegría, I. M., & Kortabarria, I. (2013). Review of wave energy technologies and the necessary power-equipment. *Renewable and sustainable energy reviews*, 27, 413-434.

Lu, S. M. (2015). *RETRACTED: A global survey of gas hydrate development and reserves: Specifically in the marine field.*

Luft, G. (2009). Dependence on Middle East energy and its impact on global security. In *Energy and environmental challenges to security* (pp. 197-210). Springer, Dordrecht.

Martišauskas, L., Augutis, J., & Krikštolaitis, R. (2018). Methodology for Energy Security Assessment considering Energy System Resilience to Disruptions. *Energy Strategy Reviews* (ELSEVIER), pp. 106-118.

Mattis, J. (2018). *Summary of the 2018 national defense strategy of the United States of America*. Department of Defense Washington United States.

Mazis, I. Th. (2008), *The Geopolitics of the wider Middle East and Turkey,* Lebanon, Athens.

McCormick, M. E. (2013). *Ocean wave energy conversion*. Courier Corporation.

Mondal, M. A. H., Ringler, C., Al-Riffai, P., Eldidi, H., Breisinger, C., & Wiebelt, M. (2019). Long-term optimization of Egypt's power sector: Policy implications. *Energy*, 166, 1063-1073.

Neumann, A. (2004). *Security of supply in liberalised European gas markets.* Diplomarbeit, European University Viadrina Frankfurt (Oder).

Paravantis, J. A., Kontoulis, N., Ballis, A., Tsirigotis, D., & Dourmas, V. (2018, July). A geopolitical review of definitions, dimensions and indicators of energy security. In *2018 9th International Conference on Information, Intelligence, Systems and Applications* (IISA) (pp. 1-8). IEEE.

Pascual, C. (2015). *The new geopolitics of energy. Center on Global Energy Policy,* Columbia University, School of International and Public Affairs. September.

Pavlović, D., Kovačić, T., Bolanča, A., & Golub, M. (2018, January). *Gas Deposits of Eastern Mediterranean-Importance and Significance in the EU Energy.* In 33. Međunarodni znanstveno-stručni susret stručnjaka za plin.

Philokyprou, M., Michael, A., Malaktou, E., & Savvides, A. (2017). Environmentally responsive design in Eastern Mediterranean. The case of vernacular architecture in the coastal, lowland and mountainous regions of Cyprus. *Building and Environment*, 111, 91-109.

Policy and Legal Documents Commission Staff Working Document: Mid-term evaluation of Council Directive 2009/119/EC. (2017, November

24). Retrieved from https://ec.europa.eu/energy/sites/ener/files/documents/mid-term_evaluation_of_council_directive_2009119ec.pdf.

Proedrou F., (2019) Contesting Greek Gas Policy: Geopolitical, Energy and Climate Considerations, *The International Spectator*, 54:2, 87-101.

Prontera, A., & Ruszel, M. (2017). Energy Security in the Eastern Mediterranean. *Middle East Policy*, 24(3), 145-162.

Renewable Energy in the Mediterranean. (2011). In C. o. t. Region (Ed.): *European Union. Reuters*. (14/01/2018). EU even more dependent on Russian gas. Business. Retrieved from https://www.thenational.ae/business/energy/eu-even-more-dependent- on-russian-gas-1.695131

Review of the European Neighbourhood Policy. (2015). *Brussels: The European Commission* Retrieved from http://eeas.europa.eu/archives/docs/enp/documents/2015/151118_joint- communication_review-of-the-enp_en.pdf.

Rutherford, J. (2018). *Natural Gas.* Retrieved from https://www.studentenergy.org/topics/natural-gas.

Sartori, N., Lorenzo, C., & Paceviciute, I. (2016). *Energy Resources and Regional Cooperation in the East Mediterranean* (No. 16, p. 27). IAI Working Papers.

Scholten, D., & Bosman, R. (2016). The geopolitics of renewables; exploring the political implications of renewable energy systems. *Technological Forecasting and Social Change*, 103, 273-283.

Shaffer, B. (2011). Israel—New natural gas producer in the Mediterranean. *Energy Policy,* 39(9), 5379-5387.

Shields, M. A., Woolf, D. K., Grist, E. P., Kerr, S. A., Jackson, A. C., Harris, R. E., & Gibb, S. W. (2011). Marine renewable energy: The ecological implications of altering the hydrodynamics of the marine environment. *Ocean & Coastal Management*, 54(1), 2-9.

Stampolis, K. (2019). *The Role of Greece in Strengthening European Energy Security*, Lecture at SETHA, Athens, 06/02/2019.

Tilliros, P. (2018). *The role of East Med Gas in the European Energy Security and the Best Cyprus Gas Monetization Option*. Retrieved from http://www.onisilos.gr/?p=17142.

Tóth, G., & Sebestyén Szép, T. (2019). Spatial Evolution of the Energy and Economic Centers of Gravity. *Resources,* 8(2), 100.

Tsakiris, Th. (2018). *Energy Security and International Politics*, Papazisis, Athens

Tsaltas, G. I., Bourtzis, T., and Rodotheatos, G. (2009) Prospects for the exploitation of submarine natural resources in the south-eastern Mediterranean with emphasis on the area between Greece - Cyprus - Turkey, 58 Presentation at the Conference of the Hellenic Army Society on: *Strategy 2009: Present Geopolitical Situation and Challenges - New Strategies Athens*, War Museum, 24 -25 November 2009.

Tsaltas, Gr. I. (2006.) *Environment and Maritime Space*, published by I, Sideri, Athens, 189-201.

Tsaltas, Gr., and Deniozos, N.(2007) *The Notion of Soft Security and the Environment in the Development of Greek-Turkish Relations in the Aegean Sea In Middle East Forum,* University of the Aegean, Issue 6, December 2007, 93-101

UfM. (2018). *L'union pour la Méditerranée - Qui sommes- nous?* Retrieved from http://ufmsecretariat.org/fr/qui-sommes-nous/

UNFICYP Fact Sheet. (2018). Retrieved from https://peacekeeping.un.org/en/mission/unficyp.

United Nations Convention on the Law of the Sea, (1982).

Urgeles, R., & Camerlenghi, A. (2013). Submarine landslides of the Mediterranean Sea: Trigger mechanisms, dynamics, and frequency-magnitude distribution. *Journal of Geophysical Research: Earth Surface,* 118(4), 2600-2618.

Vakulchuk, R., Overland, I., & Scholten, D. (2020). Renewable energy and geopolitics: A review. *Renewable and Sustainable Energy Reviews*, 109547.

Weiss, A. S., Larrabee, F. S., Bartis, J. T., & Sawak, C. A. (2012). *Promoting International Energy Security.* Volume 2: Turkey and the Caspian (No. RAND-TR-1144/2-AF). Rand Project Air Force Santa Monica CA.

Yergin, D. (2011). *The prize: The epic quest for oil, money & power.* Simon and Schuster.

Yetiv, S. A., & Lu, C. (2007). China, global energy, and the Middle East. *The Middle East Journal*, 61(2), 199-218.

Yuanita, A., Sommeng, A. N., Wijanarko, A., Yuanita, A., Migas, B. P. H., Formentini, M., ... & Anderson, R. E. (2016). The Prize: The Epic Quest for Oil Money and Power. *Asian Journal of Scientific Research,* 11(1), 1920-1933.

Index

A

B

C

D

E

R

S

T

U